MATHEMATICAL ASTRONOMY IN MEDIEVAL YEMEN

AMERICAN RESEARCH
CENTER IN EGYPT

CATALOGS / Volume 4

Published under the auspices of
THE AMERICAN RESEARCH CENTER IN EGYPT, INC.

MATHEMATICAL ASTRONOMY IN MEDIEVAL YEMEN

A Biobibliographical Survey

by

David A. King

Undena Publications
Malibu 1983

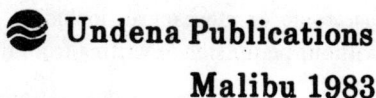

This work surveys over one hundred Yemeni astronomical manuscripts preserved in the libraries of Europe and the Near East. These sources attest to an active interest in mathematical astronomy in the Yemen from the tenth century to the early twentieth century, and the writings of various Yemeni astronomers of the thirteenth and fourteenth centuries are particularly impressive. To the historian of Islamic science some of these works are of interest because they preserve earlier Iraqi and Egyptian astronomical sources which are no longer extant in their original form, and to the historian of Islamic institutions others are of interest because they cast new light on the astronomical orientation of the Kaᶜba and on the early history of the institution of prayer in Islam.

The work is divided into two parts, the first including a survey of the history of Yemeni astronomy and classification of the sources, and the second a list of over fifty Yemeni astronomers and the available manuscripts of their works. A brief analysis of the contents of each extant work is included.

The author teaches Arabic, Islamic Studies, and History of Science at New York University. He is a specialist in medieval Arabic scientific manuscripts and has worked extensively in the manuscript libraries of Europe and the Near East. His other publications include *A Catalog of the Scientific Manuscripts in the Egyptian National Library* (Cairo, 1981) and numerous papers on different aspects of Islamic astronomy and mathematics. He is currently working on two books entitled *The World about the Kaᶜba: a Study of the Sacred Direction in Medieval Islam* and *Science in the Service of Islam.*

Library of Congress Card Number: 81-71733

ISBN: 0-89003-098-7 (paper)
0-89003-099-5 (cloth)

Undena Publications, P.O. Box 97, Malibu, CA 90265, U.S.A.

To Patricia,

Who brought me Stella and other forms of sustenance when I was stuck in quarantine at Cairo Airport for a week on my return to Egypt from the Yemen.

TABLE OF CONTENTS

LIST OF PLATES xi

ACKNOWLEDGMENTS xiii

PART I: INTRODUCTION 3

1. Introductory Remarks 3
2. Arabian Starlore and Islamic Mathematical Astronomy 3
3. Locating the Sources for the Study of Yemeni Astronomy 4
4. A Survey of the History of Yemeni Astronomy 8
5. References to the Yemeni Astronomical Tradition in Medieval Literature ... 10
6. Classification of Yemeni Astronomical Sources 11
   - A. Treatises on Folk Astronomy 11
   - B. *Zīj*es, Astronomical Handbooks 12
   - C. Tables for Astronomical Timekeeping 13
   - D. Ephemerides 13
   - E. Treatises on Instruments 14
   - F. Treatises on Astrology 14
   - G. Almanacs 14
   - H. Calendrical Tables 15

PART II: SURVEY OF YEMENI ASTRONOMERS AND THEIR WORKS 19

1. Al-Hamdānī 19
2. Ibn Raḥīq 20
3. Nashwān al-Ḥimyarī 21
4. Al-Shayzarī 22
5. Al-Aṣbaḥī 22

5a. The Sultan al-Muẓaffar 23

6. Al-Fārisī 23
7. Al-Kawāshī 27
8. The Sultan al-Ashraf ʿUmar 27
9. Abu l-ʿUqūl 30
10. The Sultan al-Muʾayyad 33
11. Anonymous (almanac) 33

11a. The Sultan al-Mujāhid 34

12. Ibn al-Mushrif 34
13. Al-Bakhāniqī 34
14. Al-Hāmilī 34
15. Al-Yāfiʿī 35
16. Anonymous (Sanaa *zīj*) 35
17. Anonymous (Taiz *zīj*) 36

18. The Sultan al-Afḍal ... 37
18a. The Sultan al-Ashraf Ismāʿīl ... 38
19. Ibn Abi l-Maʿālī ... 38
20. Al-Kaʿbī ... 38
21. Anonymous (*zīj*) ... 39
22. Anonymous (almanac) ... 39
23. Ismāʿīl al-Najrānī ... 39
24. Zayd al-Najrānī ... 40
24a. Al-Mahdī Aḥmad ibn Yaḥyā ... 40
25. Al-Hādī ʿIzz al-Dīn ... 41
26. Bā Makhrama ... 41
27. Al-Ṭawāshī ... 42
28. Al-Daylamī ... 42
29. Anonymous (planetary tables) ... 43
30. Anonymous (horoscopes) ... 43
31. Ibn Dāʿir ... 43
32. Anonymous (folk astronomy) ... 44
33. Al-Thābitī ... 44
34. Al-Dawwārī ... 44
35. Anonymous (Sanaa *zīj*) ... 45
36. Al-Ḥasan al-Sarḥī ... 45
37. ʿAbd Allāh al-Sarḥī ... 46
38. Al-Maḥallī ... 57
39. Ibn Jaḥḥāf ... 57
40. Anonymous (prayer tables) ... 47
41. Miscellaneous (calendrical tables) ... 47
42. Abu l-Qāsim al-Makkī ... 50
43. Al-Shibāmī ... 50
44. Muḥammad Ḥaydara ... 50
45. Al-Shawqānī ... 51
46. Al-ʿAnsī ... 51
47. Al-Wāsiʿī ... 51
48. Anonymous (modern prayer-tables) ... 52

APPENDIX A: YEMENI WORKS ON ARITHMETIC, INHERITANCE, AND SURVEYING ... 53

1. Al-Ṣardafī ... 53
2. Al-Ashʿarī ... 54
3. Al-Khuzāʿī ... 54
4. Al-Hāmilī ... 55
5. Aḥmad al-Jallād ... 55
6. ʿAlī al-Jallād ... 56
7. Ibn Salm ... 56
8. ʿAbd Allāh ibn ʿUmar ... 56
9. Al-Jaḥḥāf ... 56
10. Aḥmad al-Ḥusaynī ... 57
11. Al-Madhḥijī ... 57
12. Anonymous ... 57

APPENDIX B: SOME NEW MATERIAL 58

SIGLA OF MANUSCRIPTS CONSULTED 61

LIST OF MANUSCRIPTS CONSULTED 73

BIBLIOGRAPHICAL ABBREVIATIONS 75

INDEXES 81

- Index of Personal Names 81
- Index of Titles 86
- Index of Modern Authors 89
- Index of Localities 90
- General Subject Index 91
- Index of Terrestrial Latitudes 93
- Index of Values of the Obliquity of the Ecliptic 93

ADDENDA 94

CAPTIONS 95

PLATES . . . . . after p. 98

LIST OF PLATES

1. An extract from the corpus of tables for timekeeping compiled in Taiz ca. 1300 by Abu l-ʻUqūl I
2. An extract from a fourteenth-century almanac for Taiz displaying solar, lunar, and planetary positions for each day of one month of a specific year II
3. Astrological horoscopes contained in the almanac illustrated in Plate 2 III
4. An extract from a Yemeni copy of a tenth-century Iranian astronomical handbook IV
5. The back of the Sultan al-Ashraf's astrolabe preserved in the Metropolitan Museum of Art in New York V
6. The design for the back of an astrolabe illustrated in al-Ashraf's treatise on astrolabe construction VI
7. The section on the magnetic compass in the Sultan al-Ashraf's treatise on instrument construction VII
8. al-Fārisī's text on the orientation of the Kaʻba and its association with the winds VIII
9. The world arranged in sectors about the Kaʻba, as displayed in al-Fārisī's treatise on folk astronomy IX
10. An eleventh-century Yemeni poem on the Syrian months and associated celestial phenomena and agricultural activities X

LIST OF PLATES

ACKNOWLEDGMENTS

The research on medieval Islamic astronomy conducted at the American Research Center in Egypt during 1972-75 was supported by the Smithsonian Institution and National Science Foundation, Washington, D.C. Related research in the Yemen and in the Ambrosiana Library in Milan during 1973-74 was supported by a grant from the Johnson Fund of the American Philosophical Society. This support is gratefully acknowledged. I also wish to thank the directors of the various manuscript libraries in Europe and the Near East listed at the end of this study, who either allowed me access to their collections or provided me with microfilms of specific manuscripts. I am especially grateful for the privileges afforded to me in the Egyptian National Library in Cairo and the Biblioteca Ambrosiana in Milan.

It is a pleasure to thank Mr. Michael Nugent, former U.S. Cultural Attaché to the Yemen Arab Republic, for his warm hospitality during my stay in Sanaa, and Dr. and Mrs. Arnold Green for their many kindnesses during my stay in Taiz. It is likewise a pleasure to thank Qadi ʿAlī Muḥammad al-Sharafī of the Prime Minister's Office, Sanaa, for guiding me to private libraries in Sanaa, Bayt al-Faqih, Zabid, Taiz, Ibb, and Jibla. Numerous other Yemenis from different walks of life contributed to make my stay in the Yemen a most pleasant experience.

David A. King
Agami, Egypt, 1975

In preparing the final version of this monograph for publication I have mainly added references to Fuat Sezgin's latest volumes of his monumental *Geschichte des arabischen Schrifttums* and to some recent publications in the secondary literature.

It is a pleasure to thank the Orientalische Abteilung of the Deutsche Staatsbibliothek in Berlin, the Egyptian National Library in Cairo, the Biblioteca Ambrosiana in Milan, and the Metropolitan Museum of Art in New York for permission to reproduce photographs of various manuscripts and instruments in their collections.

Publication of this volume was made possible by a generous grant from the Ford Foundation to the American Research Center in Egypt. It is a pleasure to record my gratitude to the Ford Foundation for their interest in current research in the history of Islamic science.

D.A.K.
Crystal Beach, Florida, 1981

PART I

INTRODUCTION

1. Introductory Remarks

In this study I present a preliminary survey of a number of medieval astronomical manuscripts which testify to an active tradition in mathematical astronomy in the Yemen from the early tenth century to the present century. These manuscripts are of interest to the history of science and Islamic studies generally in that they constitute part of a scientific tradition that knew no rival between the eighth and fifteenth centuries, and in that they help document the history of one of the most fascinating regions in the Islamic world. Some of the manuscripts are of particular interest because they preserve material from earlier Iraqi, Egyptian, and Maghribi sources that are no longer extant in their original form; others are of particular interest because they contain material of a kind not attested in other extant Islamic astronomical sources. I am currently preparing a similar but more extensive survey of the astronomical tradition in medieval Egypt and Syria, to which the Yemeni tradition is closely related.[1]

2. Arabian Starlore and Islamic Mathematical Astronomy

A tradition of astronomical folklore has long been known to have existed in the Yemen, although it has not yet been investigated in a scholarly fashion. In the present study I am concerned mainly with mathematical astronomy, that is, planetary astronomy, spherical astronomy, and astronomical timekeeping, and also with mathematical astrology.

The pre-Islamic starlore of the Arabian peninsula relates the night sky as it changes throughout the solar year with corresponding agricultural and meteorological patterns, and is characterized by the division of the path of the moon against the background of fixed stars into twenty-eight lunar mansions.[2] This lore spread from Arabia with Islam and underlies certain simple almanacs which are known to have been used in medieval

[1] We are fortunate to have F. Sezgin's new volumes on Islamic mathematics, mathematical astronomy, astrology, and folk astronomy (*Sezgin*, V, VI, and VII in the bibliography), listing manuscripts of all known works compiled before the eleventh century, and pointing to countless areas for fruitful research. For the later period it seems to me that surveys of the material arranged chronologically according to provenance (as in *Brockelmann*) are the optimum means of dealing with the vast amounts of available Islamic scientific manuscripts. However, we no longer need mere lists of manuscripts, but rather statements of what these manuscripts contain.

I have begun a collection of notes on the astronomical traditions of Egypt and Syria in the period after that covered by Sezgin (hereafter referred to as *MAES*). For a survey of this tradition see *King* 9.

[2] On the lunar mansions see J. Ruska's article "*Manāzil*" in EI_1. See also note 3 below.

Iraq and Andalusia, and it persists to this day in such areas as Southern Arabia and the Horn of Africa. In the present study I shall to some extent neglect the Yemeni almanacs that give agricultural and meteorological information for each day of the year, some examples of which have been investigated by Prof. R. B. Serjeant of Cambridge University.[3]

Islamic mathematical astronomy, on the other hand, was ultimately derived from non-Arab sources, namely, the Indian, Sassanian, and Greek astronomical traditions.[4] It was generally not associated with traditional Arab astronomical folklore, although in the Yemen mathematical methods were applied to simple timekeeping by the lunar mansions. The mathematical tradition of Islamic astronomy is reflected in the astronomical handbooks with planetary tables called in Arabic *zīj*es,[5] and in the extensive tables for astronomical timekeeping.[6] In the medieval Yemen, as I shall show, there was considerable activity in the preparation of both *zīj*es and tables for timekeeping.

3. Locating the Sources for the Study of Yemeni Astronomy

The Yemeni astronomical manuscripts that have come to my attention are listed at the end of this study and are hereafter referred to by the appropriate sigla. There are several Yemeni manuscripts dealing with arithmetic and surveying that I have not used in this study, but which I list in an Appendix A.

My interest in Yemeni astronomy dates from 1970, when I discovered that the *Mukhtār Zīj* of Abu l-ʿUqūl, preserved in a manuscript in the British Museum (MS LA) and catalogued as Egyptian because it was based on a *zīj* of the tenth-century Cairo astronomy Ibn Yūnus, was in fact compiled in the Yemen about the year 1300. At that time I was engaged in collecting material on Ibn Yūnus, whose major work, a *zīj* dedicated to the Fatimid Caliph al-Ḥakim and called the *Ḥākimī Zīj*, is no longer extant in its entirety.[7] Routine investi-

---

[3] See the valuable study *Serjeant* 1, which contains a useful bibliography on Near Eastern folk astronomy. On the Ḥaḍramī almanac *Ḥisāb al-Shibāmī* see also Section 43.1 in Part II *Serjeant* 2, pp. 22-23, 82, and 174-175 draws attention to the astronomical knowledge of the South Arabian mariners and to the almanacs that they used.

My friend Dr. Daniel Varisco has recently embarked upon a study of medieval Yemeni almanacs and the lore relating to the lunar mansions. The results of his researches promise to be extremely interesting.

[4] The most reliable survey of early Islamic astronomy is *Pingree* 2. See also the same author's article "ʿ*Ilm al-hayʾa*" in EI_2.

[5] The standard work on Islamic *zīj*es is *Kennedy* 1, which contains a survey of over one hundred *zīj*es and an account of the contents of several notable examples. On the various categories of Islamic astronomical tables not generally contained in *zīj*es see *King* 5, and also note 6 below.

It is hoped to prepare a new survey of Islamic *zīj*es incorporating the several dozen new *zīj*es that have come to light in recent years and a critical account of the contents of all extant *zīj*es of consequence. There is no point in working on Yemeni *zīj*es until the more important Iraqi and Egyptian *zīj*es on which they were based have been investigated first; thus I have refrained from giving detailed analyses of the tables in the various Yemeni *zīj*es listed in this study.

[6] I have prepared a survey of all known Islamic tables for timekeeping by the sun and stars and for regulating the times of prayer (hereafter abbreviated *SATMI*). The survey includes an analysis of all relevant Yemeni material presently known to me (see also note 27 below).

[7] On Ibn Yūnus see *King* 1 and the article in *DSB*. I am currently preparing an analysis of Ibn Yūnus' planetary astronomy. The piecing together of the material due to Ibn Yūnus that survives only in later Egyptian and Yemeni sources is a task for the future.

gations of another anonymous fourteenth-century *zīj* (MS PA) preserved in the Bibliothèque Nationale in Paris revealed that this too was compiled in the Yemen and relied heavily on Ibn Yūnus. A century and a half ago S. Lee had drawn attention to a manuscript of the Yemeni *Muẓaffarī Zīj* compiled by Muḥammad ibn Abī Bakr al-Fārisī for the Rasulid Sultan al-Muẓaffar and preserved in Cambridge University Library (MS CZ). Upon inspecting this manuscript I found that the flyleaves were filled with tables from the *Ḥākimī Zīj*. Whilst working in the Ẓāhirīya Library in Damascus in 1970 I came across a manuscript of another Yemeni *zīj* (MS DA) based mainly on the *Muẓaffarī Zīj* but containing additional tables from the *Ḥākimī Zīj* of Ibn Yūnus and the yet earlier and no longer extant Abbasid *Mumtaḥan Zīj*.[8] It seemed worthwhile to pursue the astronomical tradition of medieval Yemen.

Obviously one would begin such an investigation with the standard bibliographical sources for Islamic astronomy. H. Suter's survey of Muslim astronomers and their works, published in 1900, contained a very few references to Yemeni astronomical works such as the no longer extant *Zīj* of al-Hamdānī, the *Maʿārij al-fikr* and *Nihayat al-idrāk* of al-Fārisī, and the *Tabṣira* of the Sultan al-Ashraf.[9] C. Brockelmann's survey of Arabic literature, of which the second edition was published between 1937 and 1949, contained a few more references to Yemeni astronomical works, notably various additional writings of al-Fārisī, Ibn Abi l-Maʿālī's poem on ephemerides, and the seventeenth-century *Zīj* of al-Muthannā al-Sarḥī.[10] E. S. Kennedy's survey of Islamic *zīj*es, published in 1956, listed only four Yemeni *zīj*es: the no longer extant *Zīj* of al-Hamdānī; the *Muẓaffarī Zīj* (known from MS CZ); the *Mukhtār Zīj* (listed as Egyptian); and the much later *Zīj al-Muthannā* (not identified as Yemeni).[11] Virtually none of these works had been studied in modern times,[12] and the only previous attempt to list Yemeni astronomers and their works, made by A. Azzawi in 1956, was based mainly on the scant information recorded by the seventeenth-century Turkish bibliographer Ḥājjī Khalīfa.[13]

[8] On the Abbasid *Mumtaḥan Zīj* see *Kennedy* 1, no. 51 and *Vernet*, and also *Pingree* 2, pp. 39-40.

[9] *Suter*, nos. 112 (al-Hamdānī, based on *al-Qifṭī*); 237 (al-Tujībī, based on *al-Maqqārī*; see note 24 below); 349 (al-Fārisī, see Section 6 below); and 394 (al-Sulṭān al-Ashraf, see Section 8 below). See also the appendix to this study.

[10] *Brockelmann*, I, p. 263 and SI, p. 409 (al-Hamdānī); I, p. 625 and SI, pp. 866-867 (al-Fārisī and al-Kawāshī; see Sections 6 and 7 in Part II); SI, p. 864 (Abu l-ʿUqūl; see Section 9 in Part II); SII, p. 253 (Ibn Abi l-Maʿālī; see Section 19 in Part II); II, p. 537 and SII, p. 567 (al-Muthannā al-Sarḥī; see Section 37 in Part II); and SII, p. 567 (al-Maḥallī, Ibn Jaḥḥāf, and Ibn Rājib; see Sections 38, 39, and 41 in Part II). See also the appendix to this study.

Brockelmann's reference (SII, p. 252) to a treatise on the astrolabe by the Rasulid Sultan al-Mujāhid is an error. He meant the treatise of al-Ashraf, and was quoting a reference by C. Nallino to MS TA (see note 12 below).

[11] See *Kennedy* 1, nos. 54 (*Muẓaffarī Zīj*), 57 (*Mukhtār Zīj*); 69 (*Zīj* of al-Hamdānī); and X212 (*Zīj al-Muthannā*).

[12] The only two previous studies of Yemeni astronomical manuscripts are *Lee* (1822), which deals with the introduction in MS CZ of the *Muẓaffarī Zīj* (see Section 6.3 in Part II), and *Jazāʾirī* (1952), which contains short extracts from MS TA of the Sultan al-Ashraf's treatise on sundials (see Section 7.2 in Part II). C. Nallino (*RSO*, 2, pp. 480-481) mentions an article on this treatise in the journal *al-Muqtabas*. I have not been able to consult this article but its title is the same as that in *Jazāʾirī*.

[13] *Azzawi*, pp. 230-235 (where MS AZ of al-Aṣbaḥī's treatise and MS TA of the Sultan al-Ashraf's treatise on the astrolabe are mentioned) and 337-338 (where al-Daylamī's *zīj* is mentioned by name only). See also the appendix to this study. Azzawi was innocent of any knowledge of any previous studies on Islamic astronomy.

Starting with the references to Yemeni astronomical manuscripts provided by Suter, Brockelmann, Kennedy, and Azzawi, I then consulted the catalogues of Arabic manuscripts in the major libraries of Europe and the Near East.[14] The largest collection of Yemeni manuscripts is preserved in the Biblioteca Ambrosiana in Milan. Excellent catalogues of this valuable collection have been prepared by E. Griffini and O. Löfgren. For other collections of Arabic manuscripts in Europe catalogues of varying quality are available, the best being Ahlwardt's catalogue of the Berlin collection. The manuscripts now in the Staatsbibliothek, Berlin that were collected by Glaser and Landberg are mainly of Yemeni provenance.

In 1971 Prof. Mahmoud Ghul of the American University in Beirut loaned to me microfilms of two astronomical manuscripts preserved in a private library in Sanaa (MSS SA and SB). The first of these was another copy of the *Muẓaffarī Zīj*, and the second contained an astronomical compendium by the Sultan al-Afḍal, rich in material from earlier Egyptian, Syrian, and Yemeni sources. In 1972 I obtained microfilms of various Yemeni astronomical manuscripts preserved in the Ambrosiana and Vatican Libraries. One of the Ambrosiana manuscripts (MS MB) contained an anonymous and incomplete set of tables for timekeeping computed for Taiz, but it was not until I was able to compare this with microfilm of a manuscript in the Staatsbibliothek in Berlin (MS BN) and the Sultan al-Afḍal's compendium from Sanaa (MS SB) that it was possible to identify the extensive corpus of timekeeping tables for Taiz entitled *Mirʾāt al-zamān* and compiled by Abu l-ʿUqūl. I was also able to locate and compare three Yemeni treatises on timekeeping dating from the eleventh (?) to the thirteenth century, namely those of Ibn Raḥīq, al-Aṣbaḥī, and al-Fārisī, preserved respectively in Berlin, Cairo, and Milan (MSS BR, TH, and ME).

In the spring of 1974 I visited several libraries in the Yemen,[15] only to find that the historical tradition investigated in this study might have been lost for all time but for the fact that a few Yemeni astronomical manuscripts came to enjoy the relative security of European libraries. However, in the Grand Mosque Library in Sanaa I came across a copy of the seventeenth-century *Zīj al-Muthannā* (MS SG) with some additional pages from a much earlier Yemeni *zīj* containing tables from the Abbasid *Mumtaḥan Zīj*. Furthermore, in Zabid I was shown a manuscript of a sixteenth-century Yemeni *zīj* called *Zād al-musāfir* (MS ZK) that contained planetary tables from the thirteenth-century Tunisian *Zīj* of Ibn Isḥāq al-Maghribī (which until 1978 was thought to be no longer extant in its original form), as well as references to another work entitled *Taysīr al-maṭālib* by an individual named al-Kawāshī. Brockelmann had noted that a manuscript of the latter exists in Alexandria Municipal Library (MS AL) and I was able to inspect this on my return to Egypt from the Yemen and to ascertain that it was another Yemeni *zīj*, based mainly on earlier Egyptian and Iraqi material. I was also able to compare photographs of an astrolabe made in 690 Hijra (= 1295/96) by the Sultan al-Ashraf and now preserved in the Metropolitan Museum of Art, New York, with his treatise on astrolabes and sundials preserved in the Egyptian National Library in Cairo (MS TA). Appended to this treatise is an account of the use of a

[14] For lists of catalogues see *Huisman* and *Sezgin*, VI. I have made extensive use of Griffini's catalogue of part of the Ambrosian collection (*Griffini* in the bibliography), and Lofgren's catalogue of the remainder (available inside the library in lithograph copy). In 1975 the first volume of a new catalogue of the Ambrosiana appeared (*Löfgren & Traini* in the bibliography). Ahlwardt's catalogue of the Berlin collection (*Ahlwardt* in the bibliography) is a fundamental reference source for any work on Islamic science.

[15] On current research facilities in the Yemen see *Green & Stuckey* and *Reinhart*.

magnetic compass to find the qibla, the earliest reference to a magnetic compass in the Islamic astronomical sources. In Cairo I also came across two Yemeni almanacs (MSS TG and TC), dating respectively from 1326 and 1405. Later in 1974 I visited the Ambrosiana Library in Milan and inspected most of the astronomical manuscripts amongst the 1,600 Yemeni manuscripts preserved in that Library, but found little new material of interest. Early in 1975 I received a microfilm of the Sultan al-Ashraf's treatise on astrology from the Bodleian Library in Oxford (MS ON). Finally, whilst preparing the final draft of this study in April, 1975, I came across a set of photographs preserved in the Egyptian National Library in Cairo of a fragment of a treatise on mathematical astrology by the early tenth-century Yemeni geographer al-Hamdānī, the original of which (MS SL) is preserved in a private library in Sanaa that I had not visited during my sojourn in the Yemen.

The vast majority of the Yemeni astronomical manuscripts now known to me are copies of Yemeni works,[16] and it appears that the only astronomical works compiled in the Yemen that were widely known and used outside that country were al-Fārisī's treatises *Maʿārij al-fikr* and *Nihāyat al-idrāk*. We also have evidence of an interest in astronomy amongst the Jews of medieval Yemen, in the form of manuscripts of al-Fārisī's *Maʿārij al-fikr* and the astrological treatise *al-Tafhīm* by the early eleventh-century Persian scholar al-Bīrūnī, all written in Arabic but in Hebrew characters.[17] It is quite clear that there were other

[16] Eight manuscripts of non-Yemeni astronomical works which either bear Yemeni notes of possession or were copied in the Yemen are the following.

First, MS YA is a thirteenth(?)-century Yemeni copy of the *Balīgh Zīj* of Kushyār ibn Labbān (*fl.* Iran, ca. 1000, cf. *Kennedy* 1, nos. 7 and 9) with later annotations by a Yemeni astronomer who attempted to adapt the planetary mean motion tables for the longitude of Sanaa.

Second, MS YE of the *Qānūn* of al-Bīrūnī (*fl.* Afghanistan, ca. 1025; cf. *Kennedy* 1, no. 59), copied in the year 1108 in Isfahan, bears a seal of ownership dated 1811 indicating that it belonged to ʿAbd Allāh ibn ʿAlī ibn al-ʿAbbās, who was *imām* of the Yemen from 1816 to 1835.

Third, MS LK is an eighteenth-century Yemeni copy of the treatise on astronomical references in the Qurʾān and the *ḥadīth* literature by the Egyptian scholar Jalāl al-Dīn al-Suyūṭī (1445-1505; cf. *Suter*, no. 449). (This treatise has recently been investigated in detail by A. Heinen of Harvard University.)

Fourth, MS MH appears to be a Yemeni copy of the Maghribi recension of the perpetual almanac of Zacuto (on whom see the article in *DSB*). It is copied in Yemeni script and bears a notice of possession of Yūsuf al-Maḥallī (see further Section 38 in Part II).

Fifth, MS BZ is an eighteenth-century Yemeni copy of the treatise on the use of the sine quadrant by Yaḥyā ibn Muḥammad al-Ḥattāb, who worked in Mecca ca. 1550 (see *Suter*, no. N 474a; *Brockelmann*, II, pp. 515-516 and SII, p. 537; and *MAES*).

Sixth, MS BG is an eighteenth-century Yemeni copy of various treatises by the early Islamic astrologers al-Ṣaymarī (*Sezgin*, VII, pp. 152-153), Sahl ibn Bishr (*Sezgin*, VII, pp. 125-128), and Māshāʾallāh (*Sezgin*, VII, pp. 102-108), and also contains a unique fragment from the astrological treatise called *al-Kafī* by Abū ʿAlī al-Marrākushī (*fl.* Cairo, ca. 1280; cf. *Suter*, no. 363).

Seventh, MS BX is a Yemeni manuscript of the treatise on theoretical astronomy entitled *Muntaha l-idrāk* by al-Kharaqī (*fl.* Marw, ca. 1125; cf. *Suter*, no. 266) copied ca. 650/1250 by Mawdūd ibn ʿUthmān ibn ʿUmar al-Mutaṭabbib al-Shirwānī.

Eighth, MS HE is a Yemeni copy of the introduction to astrology by Abu l-Qāsim al-Balkhī (cf. *Sezgin*, VII, pp. 176-177), followed by a copy of the *Zīj* of ʿAbd Allāh al-Sarḥī (see Section 37 in Part II).

Note also that MS MV_6 contains a few fragments from the *Zīj* of al-Battānī (cf. *Sezgin*, VI, p. 186, no. 1).

[17] Two such manuscripts of al-Fārisī's *Maʿārij al-fikr* are discussed in *Steinschneider* (cf. *Suter*, no. 349N). The manuscript of al-Bīrūnī's *Tafhīm*, which has been dated to ca. 1600, is discussed in *Klein-Franke*.

astronomical works of consequence compiled in the Yemen that have not survived, including not only the *Zīj* of al-Hamdānī, but also several other *zīj*es relaying on earlier Iraqi and Egyptian sources. It is not too much to hope that future research, particularly in some Near Eastern library, will turn up some of this lost material.[18]

The medieval Yemeni astronomical instruments that have survived are considerably fewer, and indeed those known to me number only two. In the courtyard of the beautiful mosque of al-Janad north of Taiz there is a stone gnomon of about 6" x 6" square cross-section and about the height of a man, which could be used to measure time of day by shadow-lengths and regulate the midday and afternoon prayers. (See Section 3.1 in Part II.) The original mosque dates from the seventh century, but I do not know when the gnomon was erected. (See also Sections 2.1 and 5.1 in Part II on timekeeping by shadow-lengths.) Aside from this gnomon, the astrolabe of the Sultan al-Ashraf preserved in the Metropolitan Museum of Art in New York (see Section 8.2 in Part II) is the only Yemeni astronomical instrument currently known to me.

4. A Survey of the History of Yemeni Astronomy

Although I have made only a preliminary examination of the sources described in the sequel, their number and variety indicate an interest in astronomy in the Yemen throughout the medieval period. This interest is characterized by an undisguised eclecticism, but with occasional originality. For the historian of science, the importance of the writings that result from such an astronomical tradition derives from the fact that much of the material incorporated by the Yemeni astronomers is no longer extant in its original form. For the historian of the Yemen, this astronomical tradition is just one new aspect of the Yemeni cultural heritage.

My survey begins with the isolated astronomical works of the celebrated tenth-century geographer al-Hạmdānī, of which unfortunately only a fragment survives. This fragment is adequate, however, to convey an impression of al-Hamdānī's familiarity with the writings of the astronomers of Abbasid Iraq.

The treatise on timekeeping by the eleventh(?)-century scholar Ibn Raḥīq is of importance for the light it casts on early Islamic timekeeping. It is representative of a class of

---

More examples of astronomical works copied in the Yemen by Jewish scholars are to be anticipated from the current researches of Prof. Bernard R. Goldstein of the University of Pittsburgh: see already *Goldstein* in the bibliography.

[18] There are, for example, several astronomical manuscripts in the Yemen that I was not able to inspect either because their owner was away from home (as in the case of the Qadi Ismāᶜīl al-Akwaᶜ and the astrologer at Bayt al-Faqīh) or because their owners refused to show them to me (as in the case of the various manuscript collections in Jibla). Whilst travelling in the Yemen I heard several times the story that the Ottoman Turks and later the Yemeni Government had confiscated large numbers of manuscripts from the owners of private collections. I was surprised not to find any astronomical manuscripts in the former library of the former Imams, now preserved in Taiz.

My friend Mr. Ayman Sayyid of Cairo kindly drew my attention to the article *Traini*, which contains a list of ninety Yemeni manuscripts preserved in Istanbul, but not one of these is a copy of a scientific work. I myself have not found any Yemeni scientific manuscripts of consequence in the major libraries there. I should be grateful for any references to collections of Yemeni manuscripts elsewhere in Turkey.

Islamic treatises that has not yet been properly studied, for no detailed account of the folk astronomy of the Arabian peninsula has been written yet.

Several manuscripts that I have come across show that the Yemeni monarchs of the Rasulid dynasty during the thirteenth and fourteenth centuries[19] not only patronized the study of astronomy but actually engaged in its study themselves. During the Rasulid period we have the extensive writings of the scholar Muḥammad ibn Abī Bakr al-Fārisī, who worked for the Sultan al-Muẓaffar and compiled for him a *zīj* for the Yemen based on the no longer extant *zīj* of the twelfth(?)-century Iraqi (?) astronomer al-Fahhād. Another astronomer who appears to have been associated with al-Muẓaffar was Muḥammad ibn Abī Bakr al-Kawāshī. He compiled a *zīj* for Taiz based mainly on earlier Egyptian and Iraqi *zīj*es that are no longer extant in their original form. The Sultan al-Ashraf, the son of al-Muẓaffar, himself compiled an extensive treatise on mathematical astrology in which he displayed his familiarity with earlier Abbasid works. His traditionalism is reflected in his use of the Indian value for the obliquity of the ecliptic. He later compiled an extensive treatise on astrolabes, sundials, and the magnetic compass, acknowledging his debt to his predecessors, the Andalusian astronomer Abu l-Ṣalt, and the Cairo astronomer Abū ʿAlī al-Marrākushī who had also written on astrolabes and sundials. In this second treatise al-Ashraf tacitly used the value for the obliquity from the *Īlkhānī Zīj*, compiled in Maragha in N.W. Persia a few decades previously.

The Sultan al-Muʾayyad, the brother of al-Ashraf, was also active in astronomy and may have been the patron of the astronomer Abu l-ʿUqūl, whose *Mukhtār Zīj* first attracted my attention to Yemeni astronomy. The *Mukhtār Zīj* appears to be based almost entirely on a *zīj* of Ibn Yūnus other than the *Ḥākimī Zīj*, and is thus of considerable importance for studies of the Fatimid astronomer. Abu l-ʿUqūl also compiled another *zīj* of which only fragments survive, but his main achievement was probably the compilation of an extensive corpus of tables for timekeeping by the sun and stars, computed for the latitude of Taiz. We may presume that Abu l-ʿUqūl was familiar with the Cairo corpus of tables for timekeeping by the sun, compiled partly by Ibn Yūnus, and mostly completed in the thirteenth century.

The Egyptian astronomer al-Bakhāniqī, who worked on the Cairo corpus in the early fourteenth century, also compiled an astronomical work in the Yemen, in fact an extension of the tables for marking astrolabes compiled by the ninth-century Abbasid astronomer al-Farghānī. The Sultan al-Afḍal, the grandson of al-Muʾayyad, prepared an extensive compendium of astronomical treatises and tables, all derived from earlier Egyptian, Syrian, and Yemeni sources.

We are fortunate to have two Yemeni almanacs from the Rasulid period, such as were doubtless prepared annually, containing planetary positions for each day of the year and extensive astrological information. Several other astronomical writings from the Rasulid period have not survived, or are only extant in fragmentary form.

In the sixteenth and seventeenth centuries *zīj*es were prepared in the Yemen by al-Daylamī, al-Najrānī, and the brothers al-Sarḥī. Al-Daylamī's *zīj* is of interest because it contains material from the *Zīj* of the thirteenth-century Tunisian astronomer Ibn Isḥāq. Al-Najrānī and the brothers al-Sarḥī relied respectively on Ibn Yūnus and al-Fārisī.

As far as I know, no astronomical works of any sophistication were compiled in the

[19] See, for example, A. S. Tritton's article "*Rasūlids*" in EI_1 for an account of the vicissitudes of their rule. There is as yet no adequate general history of the Yemen to which I can refer the reader.

Yemen after the *zīj*es of the brothers al-Sarḥī. Also, it appears that the fifteenth-century *Sulṭānī Zīj* of Ulugh Beg of Samarqand, which was adapted for the longitudes of various centres of the Ottoman Empire, was not used in the Yemen. Al-Muthannā al-Sarḥī's *Zīj* has remained popular in the Yemen until the present century, and in 1974 I met several elderly people in Sanaa and Taiz who had received instruction on it.[20] In the Yemen the last few centuries have produced a plethora of very simple calendrical tables and, apart from modern tables for regulating the times of prayer, only the simplest traditions of folk astronomy remain. Amongst the Yemenis today there is still a widespread belief in astrology and fortune telling by letter magic.[21]

5. References to the Yemeni Astronomical Tradition in Medieval Arabic Literature

I know of very few significant references in the medieval Arabic historical or biographical sources to this astronomical tradition in the Yemen or to the transmission of astronomical works to the Yemen. This is true of the scientific tradition of Islam in general: the historians and biographers were preoccupied with matters pertaining to religion and politics.

One such reference, however, is to be found in the writings of the sixteenth-century Egyptian historian Ibn Iyās,[22] who records that the rulers of the Yemen presented to the Ayyubid ruler of Egypt al-Malik al-Kāmil Muḥammad (1218-1238) an automaton (*shamaᶜdān*, literally, "candlestick") made of copper, from which each day at daybreak a brass figure of a man would step out and utter a few words of greeting to the king. Ibn Iyās added that the device had been constructed by the timekeepers (*al-mīqātīya*) and that it was lost in the days of al-Malik al-Nāṣir Muḥammad ibn Qalāwūn (1294-1341).

Another such reference occurs in the writings of the mid-fourteenth-century Egyptian historian Ibn Abi l-Faḍā'il,[23] who records that the scholar Sayf al-Dīn Abu l-Ḥasan ᶜAlī ibn Mankabris (died in 723 Hijra [= 1323]) had studied astronomy in his youth and later visited the Yemen. There he met the Sultan al-Muẓaffar and informed him of his knowledge of astronomy. He presented the Sultan with a copy of *al-Zīj al-Naṣīrī*, by which he was referring to *al-Zīj al-Īlkhānī* prepared at Maragha in N.W. Iran by the celebrated scholar Naṣīr al-Dīn al-Ṭūsī. Now Sayf al-Dīn had a debt on his shoulders, and when he made a prediction concerning the Sultan's capture of the town of Zofar that turned out to be correct, the

[20] In Sanaa I met the aged Shaykh al-Ḥātimī, called *al-Falakī*, "the astronomer," whose main interest appeared to be *ᶜilm al-zījāt*, "the science of the *zīj*es," and who carried around with him a manuscript of the *Zīj* of al-Ḥasan al-Sarḥī. My guide Qāḍī ᶜAlī al-Sharafī possessed a copy of the *Zīj* of ᶜAbd Allāh al-Sarḥī, on which he and a small group of friends had received instructions from Shaykh al-Ḥatimī. In Bayt al-Faqīh the person bearing the title *al-Falakī*, whom I unfortunately did not meet, has a nation-wide reputation as an astrologer.

[21] I am not aware of any studies of the influence of astrology or divining (*ᶜilm al-ramal*) in modern Yemeni society. A recent publication of great interest on Islamic letter magic, with a rich bibliography, is *Anawati*. On divination in medieval Islam see now the valuable study *Savage-Smith & Smith*.

[22] *Ibn Iyās*, I, p. 78. (I owe this reference to *Mayer*, p. 21.) On Islamic *automata* see the new translation of al-Jazari's treatise thereon published in *Hill* and the references there cited, and also my review in *HS*, 13 (1975), pp. 284-289.

[23] *Ibn Abi l-Faḍā'il*, p. 19 (text), p. 89 (trans.). (I owe this reference to my friend Dr. Viktoria Meinecke-Berg of the German Archaeological Institute in Cairo.)

the Sultan paid off his debt. This account illustrates the way in which a copy of the *Īlkhānī Zīj* was brought to the Yemen. It is worthy of note that al-Muẓaffar's son, the Sultan al-Ashraf, displayed his familiarity with the *Īlkhānī Zīj* in his treatise on astrolabes and sundials (see Section 7.2 in Part II).

Finally, it is worth mentioning that the early seventeenth-century Algerian historian al-Maqqārī, who spent most of his life in Syria and Egypt, writes about an Analusian astronomer named al-Tujībī who had prepared a *zīj*.[24] Al-Tujībī left Andalusia in 441 Hijra (= 1050/51) and travelled first to Egypt and then on to the Yemen. Al-Maqqārī gives no further information on the astronomer, but we can assume that in Egypt al-Tujībī had become familiar with the work of Ibn Yūnus.

There are doubtless numerous other such anecdotes and items of information in the medieval Arabic literature which have escaped my attention. Also of interest are accounts of eclipses, solar halos, and meteorites, such as are recorded by al-Khazrajī.[25]

6. Classification of Yemeni Astronomical Sources

We can divide the medieval Yemeni astronomical works that have come to light thus far into eight main categories: (A) treatises on folk astronomy; (B) *zīj*es, or handbooks containing tables for planetary and spherical astronomy; (C) tables for astronomical timekeeping; (D) ephemerides, displaying planetary positions for a given year; (E) treatises on instruments; (F) treatises on astrology; (G) almanacs; and (H) calendrical tables. I shall now list the works in these various categories, arranging them chronologically, as far as this is possible. The number in parentheses is the section in Part II where the work is discussed in greater detail. I shall then briefly survey the contents of the works attributed to each author. This survey represents the results of preliminary research only: much of the available material deserves more detailed investigation.

A. Treatises on Folk Astronomy

This group consists mainly of treatises dealing with the calendar, the determination of the qibla by non-mathematical means, and the determination of the time of day by shadow lengths and the time of night by the lunar mansions. None of these works contains any tables.

A1 Ibn Raḥīq's treatise on folk astronomy (2.1)

A2 Al-Aṣhbaḥī's *al-Yawāqīt* (5.1)

A3 Al-Fārisī's *Tuḥfat al-rāghib* (6.1)

A4 Al-Yāfiʿī's *Sirāj al-tawḥīd* (15.1)

[24] *Al-Maqqārī*, I, p. 807. (I owe this reference to *Suter*, no. 237.)

[25] *cf. al-Khazrajī*, trans., I, p. 256 and II, pp. 191, 195, 270, 280 and 281. I should be very grateful to learn from colleagues of any biographical or other material in medieval literary or historical sources which relates to the astronomical tradition in the Yemen.

A5 Al-Hāmilī's notes (14.1)

A6 Bā Makhrama's *K. al-Shāmil* (16.1)

A7 Al-Ṭawāshi's *Miftāḥ al-asrār* (27.1)

A8 Miscellaneous and anonymous (3.1, 3.2, 32.1)

B. *Zījes*

Islamic *zījes* are a clearly-defined category of astronomical handbooks containing tables for calendar conversion, planetary astronomy, and spherical astronomy, as well as instructions on their use.[26] Of the eighteen works listed below, the *zīj* of al-Hamdānī (B1) is no longer extant, and only fragments remain of the second *zīj* of Abu l-ʿUqūl (B6), the *zīj* of Ibn al-Mushrif (B7), an anonymous *zīj* for Sanaa (B8), al-Kaʿbī's *zīj* (B11), and al-Najrānī's *zīj* (B13). The astronomical compendium of the Sultan al-Afḍal (B10) is not a *zīj* but a hodgepodge of tables for planetary and spherical astronomy, and the anonymous planetary tables for Sanaa (B15) do not constitute a *zīj*. The treatise *Maʿārij al-fikr* of al-Fārisī (B3) deals with the standard topics of *zījes* but contains no tables. The remaining nine works are complete *zījes*.

B1 Al-Hamdānī's *Zīj* (1.3)

B2 Al-Fārisī's *Muẓaffarī Zīj* (6.3)

B3 Al-Fārisī's *Maʿārij al-fikr* (6.2)

B4 The *Taysīr al-maṭālib* of al-Kawāshī (7.1)

B5 Abu l-ʿUqūl's *Mukhtār Zīj* (9.1)

B6 *Zīj Abi l-ʿUqūl* (9.2)

B7 Ibn al-Mushrif's *Zīj* (12.1)

B8 Anonymous Sanaa *Zīj* (I) (16.1)

B9 Anonymous Taiz *Zīj* (17.1)

B10 The astronomical compendium of the Sultan al-Afḍal (18.1)

B11 Al-Kaʿbī's *Zīj* (20.1)

B12 Anonymous recension of the *Muẓaffarī Zīj* (21.1)

B13 Al-Najrānī's *Zīj* (23.1)

B14 Al-Daylamī's *Zād al-musāfir* (28.1)

B15 Anonymous planetary tables for Sanaa (29.1)

B16 Anonymous Sanaa *Zīj* (II) (35.1)

B17 Al-Ḥasan al-Sarḥī's *Zīj* (36.1)

B18 ʿAbd Allāh al-Sarḥī's *Zīj* (37.1)

[26] See *Kennedy* 1, pp. 139-145 for a classification of the subject matter standard in *zījes*.

C. Tables for Astronomical Timekeeping

Tables for reckoning time by the sun and stars and for regulating the astronomically defined times of Muslim prayer were generally not contained in *zījes*.[27] Only the tables of Abu l-ʿUqūl for Taiz (C2, also C3) constitute a corpus of tables for a Yemeni city. The extant tables for Sanaa (C1 and C4) are considerably less extensive. (See also Section 12.1 in Part II on what may be a fragment of a corpus for Zabid.) The other Yemeni prayer-tables that have survived are unimpressive.

C1 The tables in the Sultan al-Ashraf's astrological compendium (see F5)
C2 Abu l-ʿUqūl's corpus of tables for Taiz (9.3)
C3 The almanac of the Sultan al-Afḍal (see B10)
C4 Ibn Dāʿir's tables for Sanaa (31.1)
C5 Al-Thābitī's prayer-tables (33.1)
C6 Anonymous Ottoman-type prayer-tables (40.1)
C7 Al-Wāsiʿī's prayer-tables (47.2)
C8 Modern prayer-tables for Sanaa (48.1)
C-9 Modern prayer-tables for Taiz (48.2)

D. Ephemerides

Ephemerides, that is, tables displaying the positions of the sun, moon, and planets for each day or each few days of a given year, were compiled by Muslim astronomers from Abbasid times. From the medieval Yemen I have so far located only three ephemerides (D1, D4, and D5), the first two of which are the earliest Islamic ephemerides known to me.[28]

D1 Anonymous ephemerides for Sanaa, 727 Hijra (11.1)
D2 Anonymous notes on the compilation of ephemerides (see B10)
D3 Ibn Abi l-Maʿālī's poem on the compilation of ephemerides (19.1)
D4 Anonymous ephemerides for Taiz, 808 Hijra (20.1)
D5 Al-Maḥallī's ephemerides for Sanaa, 1146 Hijra (38.1)

[27] On the prayer-times in Islam see A. J. Wensinck's article "*Mīḳāt*" in EI_1; *King* 2, 7 and 8 which contain an analysis of the main corpuses of prayer-tables that were used in medieval Cairo, Damascus, and Istanbul; and *SATMI* which contains an analysis of all known Islamic tables for timekeeping, including a detailed discussion of all of these Yemeni tables for timekeeping (C1-C9), and also an analysis of the Yemeni treatises on timekeeping (A1-A3).

[28] Prof. B. R. Goldstein of the University of Pittsburgh has drawn my attention to two Geniza fragments in the Cambridge University Library (Taylor-Schechter 41^{103} and 42^{96}). These loose pages can be dated to 526 Hijra (= 1131-2) and 553 Hijra(?) (= 1158-9) and display astrological information for each day of the Muslim year. In the first source numerical information on the lunar crescent at the beginning of the lunar month is also presented. These fragments bear strong resemblance to the astrological parts of the Yemeni ephemerides described in Sections 11.1 and 22.1 below. (Note added in 1981: on these see now *Goldstein & Pingree*.) On the ephemerides compiled in twelfth-century Cairo see *Sayili*, pp. 167-168, quoting the early fifteenth-century historian al-Maqrīzī.

E. Treatises on Instruments

The six treatises listed below, with the exception of al-Fārisī's treatise on an eclipse computer (E1) which any way is no longer extant, deal with standard medieval Islamic instruments. The Yemeni tables for marking sundials and astrolabes (contained in E2 and E5) form part of a tradition in Islamic astronomy that goes back at least to the Abbasid astronomers al-Khwārizmī and al-Farghānī.[29]

E1 Al-Fārisī's treatise on an eclipse computer (6.5)

E2 Al-Fārisī's treatise on sundials (6.6)

E3 The Sultan al-Ashraf's treatise on astrolabes and sundials and the magnetic compass (8.2)

E4 The Sultan al-Mu᾽ayyad's treatise on the astrolabe (10.1)

E5 Al-Bakhāniqī's treatise on the astrolabe (13.1)

E6 The Sultan al-Afḍal's treatise on the armillary sphere (see B10)

F. Treatises on Astrology

A relatively small number of Yemeni scientific treatises are devoted solely to astrology, but these are of considerable interest.[30] Considerable astrological material is also contained in the Yemeni *zīj*es (see especially B5) and ephemerides (D1, D4, D5).

F1 Al-Hamdānī's *Sarā᾽ir al-ḥikma* (1.4)

F2 Al-Hamdānī's *Kitāb al-Ṭāliʿ wa-l-maṭāriḥ* (1.5)

F3 Al-Fārisī's *Nihāyat al-idrāk* (6.4)

F4 Al-Fārisī's translation of the astrological treatise of Jāmasp (6.7)

F5 The Sultan al-Ashraf's astrological compendium (8.1)

F6 Anonymous horoscopes of Yemeni rulers (30.1)

G. Almanacs

Almanacs displaying simple information relating to agriculture, meteorology, and folk astronomy for each day of the solar year, were popular in the Yemen throughout the medieval period and remain so to this day. These almanacs are of considerable interest for Yemeni folklore, and R. B. Serjeant (see note 3 above) has published a translation of one such almanac (G8) containing mainly agricultural and meteorological information.

[29] For a preliminary discussion of these tables see *King* 5, pp. 51-55 and also p. 56.

[30] Notice that the short astrological fragment published in *Khoury* is attributed to the Yemeni scholar Wahb ibn Munabbih (*fl.* ca. 700). Cf. *Sezgin*, VII, p. 99.

G1 The Sultan al-Ashraf's almanac (see F5)

G2 Abu l-ʿUqūl's almanac (9.4)

G3 Almanac for Sanaa, 727 Hijra (see D1)

G4 The Sultan al-Afḍal's almanac (see B10)

G5 Almanac for Taiz, 808 Hijra (see D4)

G6 Almanac for Sanaa, 1146 Hijra (see D5)

G7 Al-Jaḥḥāf's almanac (39.1)

G8 The *Ḥisāb al-Shibāmī* (43.1)

See also Sections 2.1 and 4.1 in Part II on the almanacs of Ibn Rahīq and al-Shayzarī.

H. Calendrical Tables

The Muslim calendar is lunar but the agricultural calendar is based on the solar year and also the times of prayer are defined in terms of the apparent daily passage of the sun across the sky. In recent centuries numerous Yemeni astronomers have complied tables for converting between the Muslim Hijra calendar and the Syrian calendar, which is based on the solar year. These calendrical tables, being so recent, are of no great interest to the history of science, and it should be remembered that most *zīj*es (especially B2 and B5 amongst the Yemeni *zīj*es) contain extensive calendrical tables that are considerably more sophisticated than the late Yemeni tables grouped below.

H1 Miscellaneous (39.2, 41.1)

H2 Al-ʿAnsī (46.1)

H3 Al-Wāsiʿī (see C7)

PART II

SURVEY OF YEMENI ASTRONOMERS AND THEIR WORKS

1. Abū Muḥammad al-Ḥasan ibn Aḥmad ibn Yaʿqūb al-Hamdānī, also called Ibn al-Ḥāʾik

Al-Hamdānī[1] is the most celebrated scholar of the medieval Yemen, renowned for his rich and varied writings on the antiquities and geography of S. Arabia and on the genealogy of its peoples, as well as for his poetry. He was born in Sanaa before the year 900 and died there about the year 950. He received his education in the Yemen, where he spent most of his life, but he also travelled extensively in the Arabian peninsula, spent some time in Mecca, and visited Iraq.

Apart from his major works on genealogy, *al-Iklīl* (see 1.1), and geography, *Ṣifat jazīrat al-ʿArab* (see 1.2), al-Hamdānī compiled at least three astronomical works (see 1.3-1.5). Of these only a fragment of a treatise on mathematical astrology survives (1.4): this fragment is of extreme importance to the history of early Islamic astronomy. We know from the thirteenth-century historian of science al-Qifṭī that al-Hamdānī also compiled a *zīj* (1.3) that was used in the Yemen until the thirteenth century: there is every reason to suppose that this *zīj* was a work of considerable sophistication.

In view of the scope of al-Hamdānī's astronomical works and the fact that this astronomical tables were apparently used in the Yemen for about three centuries, I find it surprising that there is not a single reference to him in any of the later Yemeni astronomical writings that I have examined.

1.1 al-Iklīl

Whilst there is no astronomical material in the surviving portions of al-Hamdānī's encyclopaedic work *al-Iklīl*, al-Qifṭī states that it contained considerable information on planetary conjunctions and astrology, as well as the physical sciences, and mentions that al-Hamdānī recorded the statements of his predecessors on the creation of the universe and their differing opinions on the motions of its constituent parts.[2]

[1] On al-Hamdānī see *Suter*, no. 112; *Brockelmann*, I, pp. 263-264, and SI, p. 409; *Kennedy* 1, no. 69; *Sezgin*, II, p. 650 and VII, pp. 164-165 and 272-273; *Sayyid*, pp. 68-76; and the articles in EI_2 by O. Löfgren and in *DSB* by C. Toll. The Alexandria manuscript of the *Risāla fī l-Ghālib wa-l-maghlūb* mentioned by Brockelmann (I, p. 263) is unrelated to al-Hamdānī.

[2] *Al-Qifṭī*, p. 163.

1.2 Ṣifat jazīrat al-ʿArab

In his book on the geography of the Arabian peninsula al-Hamdānī first discusses the seven climates, giving for each the latitudes and shadows at the equinoxes and solstices. The latitudes are as given in Ptolemy's *Almagest*, and the shadows differ from the *Almagest* values in that they are expressed in terms of gnomon length 12 rather than Ptolemy's 60.[3]

1.3 *Zīj* (B1)

In the surviving fragment of his treaties on mathematical astrology, the *Sarāʾir al-ḥikma* (see 1.4), al-Hamdānī mentions that he compiled a *zīj* but gives no information whatsoever on its contents. Al-Qifṭī also records that al-Hamdanı compiled a *zīj*, and implies that it was still in use in the Yemen in the thirteenth century (Arabic, *ʿalayhi ʿtimād ahl al-yaman*). Unfortunately this work is no longer extant. Since al-Hamdānī is known to have been familiar with the works of the early Abbasid astronomers, his *Zīj* would no doubt be of great interest to us. (See 16.1 below for a fragment of an early Yemeni *zīj* containing material from the Abbasid *Mumtaḥan Zīj*.)

1.4 Kitāb Sarāʾir al-ḥikma (F1)

MS SL, copied in an untidy Yemeni hand, is the sole surviving fragment of an extensive treatise by al-Hamdānī dealing with mathematical astrology. The fragment contains only the tenth *maqāla* of this work, and this *maqāla* alone, divided into 33 *bāb*s, takes up 32 folios of manuscript. A detailed investigation of this manuscript at the hands of a specialist would constitute an important contribution to our knowledge of early Islamic astronomy and astrology, because al-Hamdānī mentions the opinions of several earlier astronomers from the eighth and ninth centuries. A list of titles of the chapters has been published by F. Sezgin.[4]

1.5 Kitāb al-Ṭāliʿ wa-l-maṭāriḥ (F2)

Al-Qifṭī mentions a work with this title by al-Hamdānī, the subject of which was probably mathematical astrology since the words *ṭāliʿ* and *maṭāriḥ* refer to the horoscopus and the projections of the rays, respectively.

2. Abū ʿAbd Allāh Muḥammad ibn Raḥīq ibn ʿAbd al-Karīm

This individual is unknown to the modern literature. I agree with Ahlwardt's dating him

[3] Cf. *al-Hamdānī* 1, especially II, p. 5 and the foldout between pp. 240-241, on which al-Hamdānī's values are compared with those of Ptolemy.

[4] Cf. *Sezgin*, VII, pp. 164-165.

to the early eleventh century.[5] From the treatise described below it is clear that he worked in Mecca, but was in contact with the folk-astronomical traditions in Yemen and Egypt.

2.1 Treatise on folk astronomy (A1)

MS BR, preserved in Berlin, is the only known copy of a treatise on folk astronomy by Ibn Raḥīq. The manuscript contains 71 folios out of an original manuscript of 88 folios, and can be dated to the thirteenth or fourteenth century. The author's name is clearly written on the title and folio. The beginning of the title has been torn away: the title ends . . . *ʿalā madhāhib al-ʿArab*. A notice of possession on the title folio reads: *intaqala hādha l-kitāb al-manāzil (sic) ila l-ʿabd al-faqīr . . . ʿAlī ibn Ḥamza* . . ., so that the original title may have been *Kitāb al-Manāzil . . . (?) ʿalā madhāhib al-ʿArab*.

The first part of Ibn Raḥīq's treatise deals with the calendar, the lunar mansions, the qibla, and timekeeping. The second part (fol. 39 ff.) deals with the characteristics of the lunar mansions, giving such information as the dates when dawn breaks and the sun is in each mansion and the corresponding shadow lengths at Mecca. Some of the material on timekeeping is said to be taken from a book by Abū ʿAlī ʿArafa, the muezzin of the mosque of ʿAmr in Fusṭāṭ (*al-jāmiʿ al-ʿatīq*) (see, for example, fols. 12v and 21r). Likewise some of the material on the *qibla* is said to be taken from the *Kitāb Dalāʾil al-qibla* by Muḥammad ibn Surāqa al-ʿĀmirī (who died in 410/1019). Ibn Raḥīq also mentions Aḥmad al-Mʿd (?) Aḥmad al-Rʾry (al-Rāzī?), the muezzin at the Sacred Mosque (*al-masjid al-ḥarām*) in Mecca (fol. 12r), Maʿshuq al-Qararī, author of "*al-kitāb al-kabīr*" (fol. 39r), and Abū Jaʿfar al-Rāsibī (fol. 14r), and Aḥmad ibn Ṣalāḥ and ʿAbd al-Rāziq (or al-Razzāq), both of Sanaa (fol. 13r), as well as various poets and transmitters of *ḥadīth*. The reference to the two Yemenis is quoted from Abū ʿAlī ʿArafa and informs us that the former measured the meridian shadow in Sanaa in the presence of the latter, who was a *muḥaddith*, on Kayhak 25 of an unspecified year, and found it to be 4 ½ units.

This treatise deserves detailed study. I have already used some of Ibn Raḥīq's information on timekeeping by shadow lengths to explain the origin of the definitions of the times of the daytime prayers in Islam,[6] and elsewhere have abstracted the material on timekeeping[7] and the determination of the qibla by non-technical means.[8]

3. Nashwān Ibn Saʿīd ibn Salāma al-Ḥimyarī

A celebrated poet, historian, and grammarian, who died in 573/1177.[9]

[5] See *Ahlwardt*, pp. 149-150, where a list of chapter headings is given. The existence of Ibn Raḥīq's treatise is noted in *Brockelmann* I, p. 257.

[6] See *King* 6, Appendix B on pp. 193-196: "A note on the definitions of the times for the daytime prayers in Islam," and *SATMI*, IV.

[7] See *SATMI*, II and III.

[8] See *King* 11, Part 3.

[9] Cf. *Brockelmann*, I, p. 364, and SI, pp. 527-528; and *Sayyid*, pp. 78-79, and 110-111. Brockelmann lists various other copies of his poem in the Vatican and Ambrosiana Libraries.

3.1 Shams al-ʿulūm fī dhikr al-kabīsa (A8)

Perhaps a treatise on the calendar, extant in several manuscripts: not inspected.

3.2 Urjūza fi l-shuhūr al-Rūmīya (A8)

MS YK contains a poem on the Syrian months by Nashwān. See also MS VK.

4. Abu l-Ghanāʾim Muslim ibn Maḥmūd al-Shayzarī

Brockelmann notes two non-scientific works by this individual, who worked for the last Ayyubid ruler of the Yemen al-Malik al-Masʿūd Ṣalāḥ al-Dīn Yūsuf (d. 626/1229).[10]

4.1 ʿĀdāt al-nujūm

MS SM, copied in 1076/1665-66 and preserved in Sanaa, is an apparently unique copy of an almanac by al-Shayzarī which closely resembles the celebrated *Calendar of Cordova*.[11] The astronomical information contained in al-Shayzarī's almanac, such as the solar meridian altitude (52½° at the equinoxes) and midday shadow lengths for each month, indicates that the work was *not* compiled for use in the Yemen.

5. Ibrahīm ibn ʿAlī ibn Muḥammad al-Janadī al-Aṣbaḥī [12]

5.1 Kitāb al-Yawāqīt fī ʿilm al-mawāqīt (A2)

MS AZ, copied in the Taiz in 680 Hijra (= 1281-82), and MSS YF, YG, and TH_1, are copies of a work on timekeeping bearing this title. I have not been able to consult MS AZ, preserved in Baghdad, and have relied on the other three later copies. This work is a particularly valuable source of information on early Islamic practice. It is arranged in numerous unnumbered *faṣl*s, and al-Aṣbaḥī[12] quotes from the *Kitāb al-Mawāqīt* of Abū Jaʿfar al-Baṣrī (ca. 270 Hijra) and the *Kitāb Dalāʾil al-qibla* of Abu l-ʿAbbās ibn al-Qāṣṣ.[12a] The work consists mainly of quotations from the *ḥadīth* and the early legal scholars concerning the times of prayer and the calendar, the *qibla*, and crescent visibility.[13] The author concludes his work by saying that any such book should end with a section on the shadows in the place where the book was written, and so the book ends with a set of midday shadows for each

[10] *Brockelmann*, I, p. 302 and SI, p. 306.

[11] Cf. *Dozy & Pellat* for the text and translation, and *SATMI*, III for comments on the astronomical content.

[12] Al-Aṣbaḥī is not mentioned in the modern sources, other than *Azzawi*, p. 231.

[12a] Cf. *Sezgin*, I, pp. 496-497. This treatise is investigated in *King* 11.

[13] See further *SATMI*, IV, which includes various quotations from al-Aṣbaḥī's treatise relating to the times of prayer.

thirteen days of the year, which the author says he observed after finishing the work in the mosque of al-Janad in 654 Hijra (= 1256).

5a. Al-Sulṭān al-Muẓaffar Yūsuf ibn ʿUmar

He was Sultan of the Yemen from ca. 1250 to 1295, and was the patron of al-Fārisī (see 6). Ḥājjī Khalīfa attributes al-Kawāshī's *Zīj* to him (see 7.1).

6. Abū ʿAbd Allāh Muḥammad ibn Abī Bakr ibn Muḥammad ibn Abī Bakr ibn Ḥasan ibn ʿAlī al-Fārisī al-Tammī(?)[14]

Al-Fārisī was a scholar of wide learning, author of several works on astronomy, medicine, music, and magic. His father had emigrated from Persia (hence the name al-Fārisī), but al-Fārisī himself was born in Aden. He later worked in the service of the Rasulid Sultan al-Muẓaffar and died in 677 Hijra (= 1278/9). This information is recorded by the late fourteenth-century Yemeni historian al-Khazrajī,[15] who also notes the titles of some of al-Fārisī's works on medicine, music, and magic, but not his astronomical works.

Al-Fārisī's *Zīj* (see 6.3 below) was widely used in the Yemen for several centuries and was adapted by later Yemeni astronomers. His treatises on the subject matter of *zīj*es (see 6.2 below) and on astrology (see 6.4 below) were both known outside the Yemen as well. The modern biobibliographical sources on al-Fārisī are very confused.[16]

6.1 Tuḥfat al-rāghib wa-turfat al-ṭālib fī taysīr al-nayyirayn wa-ḥarakāt al-kawākib (A3)

MS ME is a copy of an unsophisticated treatise on astronomy with this title by al-Fārisī. MS BM is an anonymous fragment from the same work. The work is divided into twelve

[14] Al-Fārisī's name is given thus by al-Khazrajī (see note 2). Brockelmann (see note 3) gives the *kunya* Badr al-Din, which I have not noticed in any manuscripts. In fact, most of the astronomical manuscripts have simply Muḥammad ibn Abī Bakr al-Fārisī.

[15] *Al-Khazrajī*, I, p. 204.

[16] Ḥājjī Khalīfa mentions al-Fārisī's *Zīj* and his treatise *Nihāyat al-idrāk* (*HK*, II, p. 1985). Elsewhere (II, p. 1855) he states that Muḥammad ibn Abī Bakr al-Fārisí died in 629 Hijra (1231-32), which date has penetrated the bibliographical literature on the Yemeni astronomer. The date seems secure: in Flügel's edition of Ḥājjī Khalīfa's work it is written in numbers, and in the 1941 Istanbul edition it is written in words. Ḥājjī Khalīfa's statement, which is taken from a work entitled *Muntaha l-suʾl wa-l-aml fī ʿilmayi l-uṣūl wa-l-jadl* by Ibn al-Ḥājib al-Malikī (d. 646 Hijra, cf. *GAL*, I, pp. 367-373 and SI, pp. 531-539), does not refer to the Yemeni astronomer. Also, the colophons of some of the manuscripts of the *Nihāyat al-idrāk* (see 5.4) state that the work was compiled in 606 Hijra (= 1209). The problem of these dates was discussed by *Suter*, p. 218 (note 72) and no. 349N. See further note 24 below.

Suter (no. 349 and 349N) lists the *Nihayāt al-idrāk* and *Maʿārij al-fikr*, and was aware of the problem of al-Fārisī's dates (see note 2).

bābs,[17] and contains simple rules for finding the ecliptic positions of the sun and moon for a given date, and for reckoning twilight by the lunar mansions, and also contains a discussion of the midday shadow lengths throughout the year, lunar crescent visibility, and the determination of the qibla using the sun, stars, and winds.[18] A particularly interesting section of the treatise deals with the astronomical orientation of the Kaʿba.[19] In the discussion of the qibla the town of Aden is specifically mentioned (fol. 22r of MS ME), and in a section on calendars there is also a reference (fol. 23r) to Kūshyār (ibn Labbān) al-Jīlī, an astronomer who worked in Iran ca. 1000.[20] The subject matter of the work is similar to in the treatise by the Hijazi scholar Muḥammad ibn Raḥīq (see 2.1). Another historical reference is a worked example on the equation of time stated to be taken from the *Zīj* of Muḥammad ibn Yaḥya ibn (Abī) Manṣūr, whose father compiled the *Mumtaḥan Zīj* for the Abbasid Caliph al-Maʾmūn.[21] The final chapter of al-Fārisī's treatise deals with the determination of the *qibla* and contains a worked example for Aden (latitude 13°).

6.2 Maʿārij al-fikr al-wahīj fī ḥall mushkilāt al-zīj (B3)

MSS AR, JA_1, NE, RA, TF, and TG_1 are copies of another treatise by al-Fārisī that is more sophisticated than his *Tuḥfat al-rāghib* (see 6.1), being a discussion of the standard topics of planetary and spherical astronomy dealt with in the introductions of *zīj*es. The treatment is less technical than that of most *zīj*es and the text is illustrated with diagrams of the Ptolemaic planetary models and simple configurations illustrating problems in spherical astronomy. Al-Fārisī pays tribute to the Sultan al-Muẓaffar in his introduction and states that his treatise was compiled for the royal treasury (see Section 6.3).

Brockelmann (I, p. 625 and SI, pp. 866-867) neglected to mention al-Fārisī's *Zīj* and identified al-Kawāshī with al-Fārisī. He also confused the authorship of the medical work *Kitāb al-Durra al-muntakhaba* (cf. SI, p. 867 and SII, p. 252). Ullmann (1, pp. 312 and 340) states that Raḍi l-Dīn Abū Bakr ibn Muḥammad al-Fārisī, author of the *Kitāb al-Durra al-muntakhaba* . . . , was the son of Badr al-Dīn Muḥammad ibn Abī Bakr al-Fārisī, the Yemeni astronomer. Ḥājjī Khalīfa (*HK*, I, p. 744) attributes this title to an individual called Naṣr ibn Naṣr, and states that it was dedicated to Dāʾūd, the son of al-Malik Manṣur.

Azzawi (pp. 232-233) got tied up with al-Fārisī and the Sultan al-Muẓaffar. He noted the problem of the date 627 Hijra for al-Fārisī's death but not of the date 606 Hijra for the compilation of the *Nihāyat al-idrāk*.

Mayer (p. 59) confused the maker of the Oxford eclipse computer with the Yemeni astronomer. Ullmann (2, p. 342) mentioned al-Fārisī's astrological works and correctly associated the astronomer with the Sultan al-Muẓaffar, thereby, however, overlooking the date on the colophon of the *Nihāyat al-idrāk* (on which see note 24 below).

[17] See *Ahlwardt*, p. 190 for a table of contents of this manuscript.

[18] On the shadow schemes see *SATMI*, III. On the material relating to the qibla see *King* 11, Section 3.

[19] This is published with translation in *Hawkins & King*.

[20] On Kūshyār and his astronomical works see *Sezgin*, VI, pp. 246-249.

[21] On the *Mumtaḥan Zīj* see *Kennedy* 1, no. 51. I know of no other references to a *zīj* by Yaḥyā's son.

6.3 Al-Zīj al-Mumtaḥan al-Khazā'inī = al-Zīj al-Muẓaffarī = Zīj al-Fārisī (B2)

The *zīj* prepared for the Sultan al-Muẓaffar by al-Fārisī is extant in MSS CZ_2 and SF. It consists of an introduction of forty chapters and extensive tables.[22] MS SA is a third copy of the *zīj* which lacks the introduction.

In his introduction al-Fārisī states that his planetary tables, which are computed for the Persian calendar, are based on the parameters of al-Fahhād derived (in Iraq?) a century previously. The *Muẓaffarī Zīj* is in fact based on the *ʿAlā'ī Zīj* of al-Fahhād, a work which was also used by Byzantine astronomers.[23] Since none of the al-Fahhād's six *zīj*es have survived, the *Muẓaffarī Zīj* is an important source for our knowledge of the tradition of al-Fahhād. The introduction to the *Muẓaffarī Zīj* is of particular historical interest because al-Fārisī makes critical mention of no less than twenty-eight earlier *zīj*es. This passage was probably lifted from al-Fahhād. The spherical astronomical tables in the *Muẓaffarī Zīj* involving terrestrial latitude are based on latitude 14;30°. These were probably computed by al-Fārisī himself.

An anonymous Yemeni astronomer prepared a recension of al-Fārisī's *Zīj* about the year 1400, in which he converted the planetary tables to the Hijra calendar (see 21.1).

6.4 Risālat Nihāyat al-idrāk fī asrār ʿulūm al-aflāk (F3)

MSS DP, NP, NQ, TI (seven copies), TJ, YC, YD, and YQ are copies of an astrological treatise which al-Fārisī compiled for the treasury of the Sultan al-Muẓaffar. Most if not all of these manuscripts are not of Yemeni provenance and date from the seventeenth century onwards. In the introduction al-Fārisī mentions that he had also written three other works for the treasury, two of which dealt with sundials (see Section 6.6) and an eclipse computer (see Section 6.5), and the third of which dealt with musical theory and instruments. The *Nihāyat al-idrāk* is divided into three sections (*maqṣad*), the first two dealing with *ikhtiyārāt* and the third with the twelve astrological houses. The date of compilation is given in most manuscripts as 606 Hijra, which is false.[24]

6.5 Al-Risāla al-Muẓaffarīya fī l-ʿamal [bi-l-āla?] al-musammā bi-l-ṣafīḥa al-jawzaharīya (E1)

Al-Fārisī mentions this title in his introduction to the *Nihāyat al-idrāk* (see 5.4), adding that he invented the instrument called *ṣafīḥa jawzaharīya*, "the plate of the nodes," for determining lunar eclipses. No copies of his treatise are known to me.

[22] The introduction to the *Muẓaffarī Zīj* was first studied in *Lee* (1822) using MS CA. See also *Kennedy* 1, no. 54.

[23] On the *zīj*es of al-Fahhād see *Kennedy* 1, nos. 23, 53, 58, 62, 64, and 84, and on the Byzantine tradition see *Pingree* 1.

[24] The dates of compilation given in the colophons of the various copies of this work are garbled. The various dates given are Saturday, Rabīʿ I 12 or 13 or 22 or 23, 606 Hijra, and these dates are written out in words in all cases. There are too many variables to try to determine the correct date with any certitude. However, one possibility would be Rabīʿ I 12, 660 Hijra (= February 4, 1262), which was a Saturday.

In view of the rarity of references to eclipse computers in the Islamic sources[25] it is a remarkable coincidence that the only surviving Islamic eclipse computer, now preserved in the Museum of the History of Science in Oxford,[26] was made by an individual called Muḥammad ibn Abī Bakr ibn Muḥammad al-Rashīdī al-Ibarī al-Iṣfahānī. This instrument is dated 618 Hijra (= 1221/2) and bears plates for various latitudes between 30° and 40°. It seems not unlikely that the Yemeni astronomer al-Fārisī was related to the earlier instrument maker, and was perhaps his grandson. However al-Fārisī's great-great-grandfather's name, Ḥasan, given by al-Khazrajī, is not the same as the name of the grandfather of the maker of the Oxford instrument, Muḥammad. In view of the fact that Ḥājjī Khalīfa's Muḥammad ibn Abī Bakr al-Fārisī died about ten years after the Oxford instrument was made, it is not surprising that the Oxford instrument has been attributed to the Yemeni astronomer himself.

6.6 Al-Risāla al-Ẓillīya (= Risālat al-Ẓill al-mabsūṭ) (E2)

Al-Fārisī mentions this treatise on sundials in his introduction to the *Nihāyat al-idrāk* (see 6.4), stating that it dealt with "the marking of the instruments with lines for finding the hour of day and the times of the prayers, and which dispense with all other instruments." This treatise is not extant, and the only other known Yemeni treatise on sundials is that compiled by the Sultan al-Ashraf (see 8.2 below).

6.7 al-Aḥkām al-Jāmaspīya fī asrār ḥarakāt al-abwāb al-ʿulwīya = Ṭirāz al-dahr fī asrār al-khalq wa-l-amr fī l-aḥkām al-Jāmaspīya (F4)

MS MV_3 is the only known copy of an Arabic translation of an astrological treatise of Jāmasp, a contemporary of Zarathustra, made by al-Fārisī.[27] The translator states in the introduction that his father owned a copy of the Pahlavi text of this work but that he had been stingy (*ḍanīn*) with it. However, upon his father's death, he had studied it and then translated it. The treatise deals with the Flood and Noah; the coming of Abraham and Moses, Alexander, Christ, and spread of Christianity, the Jewish state and the Roman Empire; the coming of Mani; the Brahmans, Bahram, and Mazdak; and the coming of Ibn Hāshim and the Prophet Muḥammad.

[25] The only other known Muslim astronomers who wrote on this kind of instrument are three: firstly ʿAlī ibn ʿĪsā (*fl.* ca. 830, cf. *Krause*, no. 23), secondly al-Bīrūnī (*fl.* Ghazna, ca. 1025, discussed in *Wiedemann*), and thirdly al-Marrākushī (*fl.* Cairo, ca. 1280, based on al-Bīrūnī, discussed in *Sédillot-fils*, pp. 206-209). Planetary equatoria can also be used for determining eclipses. See, for example, the treatise of al-Kāshī (*fl.* Samarqand, ca. 1425), discussed in *Kennedy* 2, pp. 222-236.

[26] On the instrument see further *Gunther*, I, pp. 118-120; *Mayer*, p. 59; and *Price*, pp. 54-56.

[27] On the words of Jāmasp in the Arabic sources see *Ullmann* 2, pp. 295-296, and 342, and *Sezgin*, VII, pp. 86-88.

7. Muḥammad ibn Abī Bakr al-Kawāshī

Since the *zīj* attributed to al-Kawāshī contains material compiled at the earliest in 1284, and since Muḥammad ibn Abī Bakr al-Fārisī (see 6 above) died in 1278/79, al-Kawāshī cannot be identical with al-Fārisī.[28]

7.1 Taysīr al-maṭālib fī tasyīr al-kawākib (B4)

MS AL, preserved in Alexandria, contains a *zīj* with this title attributed to Muḥammad ibn Abī Bakr al-Kawāshī, copied in Sanaa by Yūsuf ibn Yūsuf al-Maḥallī (see 38 below) in 1142/1730. Ḥājjī Khalīfa mentions the same title and records the same *incipit* that is found in MS AL, but he attributes the work to Abu l-Manṣūr Yūsuf ibn ʿUmar, the Rasulid Sultan al-Muẓaffar (see 5a above).[29] MS LM, preserved in the British Library in London, is a second copy of this *zīj*, perhaps dating from about 800/1400, but it lacks the title folio and bears no author's name. In MS ZK_1 of al-Daylamī's *Zīj* (see 28.1 below) certain tables are said to have been taken from the *Taysīr al-maṭālib* of al-Kawāshī.

The *Taysīr al-maṭālib* is based partly on an earlier *zīj*, as yet unidentified but of Iraqi provenance and probably dating from Abbasid times. It also contains the solar and lunar equation tables of Ibn Yūnus. The planetary mean motion tables are based on the Hijra calendar, and are said to have been compiled for the longitude of Taiz. The work contains a table lifted from the Abbasid *zīj*, which displays the *qibla* as a function of geographical latitude and longitude.[30] It also contains tables of certain standard spherical astronomical functions for the latitudes of Aden (11;0°) and Taiz (13;40°), based on obliquity 23;33°, a parameter usually associated with the Abbasid *Mumtaḥan Zīj* but here intended as an approximation for 23;32,50°, a distinctive parameter which al-Kawāshī states he found by observation. Of considerable interest is a list of thirteen planetary observations made at Qus in the Nile Valley and at Alexandria between 1273 and 1284, recorded in the introduction.[31] Curiously, al-Kawāshī is not mentioned in any of the medieval Egyptian sources known to me. The date 1284 serves as a *terminus post quem* for the compilation of the *Taysīr al-maṭālib*, a work which deserves detailed study.

8. Al-Sulṭān al-Ashraf ʿUmar ibn Yūsuf

Son of the Sultan al-Muẓaffar, and Sultan of the Yemen from 1295 to 1297.[32]

[28] The reading of al-Kawāshī's name is not certain. In MS AL of his *zīj*, the name is written *al-kwʾs*, which can be read al-Kawwās or al-Kawwāsh, which means "baker" (*Dozy*, II, p. 499b). The existence of MS AL was noted by Brockelmann (I, p. 625), who, not unreasonably, identified "al-Kauwās" with al-Fārisī. Al-Kawāshī as a *nisba* is attested elsewhere (cf. *Brockelmann*, I, p. 529) and relates to the town of al-Kawāsha in the mountains east of Mosul.

[29] *HK*, I, p. 519.

[30] On this same *qibla* table in other medieval Islamic sources see *King* 4, Appendix C.

[31] An analysis of these observations is contained in *Gingerich & King*.

[32] On al-Ashraf and his works see *Suter*, no. 394; *Brockelmann*, I, p. 650, and SI, p. 901; *Azzawi*, pp. 233-234; *Sayyid*, pp. 131-132; *Ullmann* 2, p. 342 and the references there cited.

8.1 Kitāb al-Tabṣira fī ʿilm al-nujūm (F5, also C1 and G1)

MS ON contains an extensive astrological compendium with this title by the Sultan al-Ashraf.[33] Most of the chapters deal with astrology and there are occasional references to Dorotheus, Ibn Nawbakht, Sinān ibn al-Fatḥ al-Ḥarrānī, and Kūshyār. However, the compendium also contains lengthy sections on timekeeping, including tables displaying the solar altitude and longitude of the horoscopus as functions of the solar longitude for each seasonal hour of the day. Other tables display the solar altitude at midday and at the beginning of the afternoon prayer, as well as the corresponding longitude of the horoscopus, for each degree of solar longitude. The underlying parameters are latitude 14;30° (Sanaa) and obliquity 24°. The table of shadows (cotangents) is based on a gnomon length of 6;40. The table of geographical coordinates includes the following latitudes: Sanaa, 14;30°; Aden, 13;0°; Zabid, 14;0°, and Taiz, 13;43°. This last value is not attested in any other known source; note that al-Ashraf used 13;37° for Taiz (with obliquity 23;30°) in his treatise on the astrolabe (see 8.2). In addition to this material al-Ashraf presents considerable information on the lunar mansions and the constellations, and his treatise also contains an almanac.

8.2 Treatise on astrolabe and sundials (E3)

MS TA, apparently copied about 690 Hijra (= 1291), and MS TZ, copied in 888 Hijra (= 1483/84), are two copies of an extensive treatise on the construction of the astrolabe and horizontal sundial by al-Ashraf.[34] I have not inspected MS TZ. In MS TA the title is given as *Muʿīn al-ṭālib fi l-ʿamal bi-l-asṭurlāb*, "The student's aid on the use of the astrolabe," and in MS TZ it is given as *Manhaj al-ṭullāb fi l-ʿamal bi-l-asṭurlāb*, "The students' course on the use of the astrolabe." Both of these titles may be spurious.

In the first part of his treatise dealing with the construction of the astrolabe, al-Ashraf refers to the treatise on the use of the astrolabe by Abu l-Ṣalt Umayya ibn Abi l-Ṣalt (MS TA, fol. 20r), a late eleventh-century Andalusian astronomer whose treatise was written whilst he was in prison in Cairo,[35] and also mentions the extensive treatise on spherical astronomy and instruments by the late thirteenth-century Cairo astronomer Abū ʿAlī al-Marrākushī.[36] He clearly had access to some other astronomical works which are no longer extant: of particular interest is his remark that the Fatimid Caliph al-Ḥākim, (the patron of Ibn Yūnus), had an armillary sphere consisting of nine rings each of which

[33] The *Tabṣira* is mentioned by Ḥajjī Khalīfa (*HK*, I, p. 338) and the Oxford manuscript thereof is listed in *Suter*, *Brockelmann*, and *Ullmann* 2.

[34] The brief article *Jazʾāirī* (mentioned in *Azzawi*, p. 234) notes the existence of the Cairo manuscript (MS TA). Otherwise this treatise is not mentioned in the modern sources. The article also contains the text of al-Ashraf's description of the instruments of al-Ḥākim and Hulagu, as well as the text of the certificates for al-Ashraf's astrolabes.

[35] On Abu l-Ṣalt see *Suter*, no. 272.

[36] On al-Marrākushī see *Suter*, no. 363; *Krause*, no. 363, *Ullmann* 2, p. 341, and *MAES*. The first part of his treatise dealing with spherical astronomy and sundials was translated in *Sédillot-père*. The second part dealing with astronomical instruments was summarized in *Sédillot-fils*.

weighed 2,000 *ruṭl*s (about 2,000 pounds) and was large enough for a man to ride through on horseback.[37]

The text is illustrated with diagrams, and tables of coordinates for marking the almucantar and azimuth curves on astrolabe plates are given for latitudes

13°, 13;37°, 14°, 14;30°, 15;0°, 21°, and 24°.

Note the distinctive value 13;37°, which is intended to be the latitude of Taiz (MS TA, fols. 10v and 127v, cf. 9.2 and 17.1). These tables are similar in conception to those of the ninth-century Baghdad astronomer al-Farghānī, and al-Ashraf also reproduces al-Farghānī's auxiliary table (based on obliquity 23;33°).[38] Al-Marrākushī had included the same table in his treatise, attributing it to "one of the predecessors."

Al-Ashraf next deals with horizontal sundials, presenting tables of coordinates for marking the seasonal hours on the shadow traces of the zodiacal signs, computed for each of the latitudes mentioned above. The underlying value of the obliquity is 23;30° (MS TA, fol. 7r), which al-Ashraf probably adopted from the *Īlkhānī Zīj* of Naṣīr al-Dīn al-Ṭūsī, whose patron Hulagu is mentioned in al-Ashraf's introduction.[39] The sundial tables are of the same kind as those of al-Khwārizmī (using 23;51° for the obliquity)[40] and al-Marrākushī (using 23;35° for the obliquity).[41]

Al-Ashraf's work ends with a treatise on the use of the magnetic compass (*ṭāsa*) to find the qibla. This treatise is of considerable historical interest as the earliest attested reference to the compass in an Islamic astronomical source.[42] Al-Ashraf mentions the direction 27° west of north as the qibla, adding that for central Yemen, in which he includes Aden, Taiz, and Zabid, the qibla is 29° west of north (MS TA, fol. 145r).

Following al-Ashraf's treatise in MS TA are two *ijāza*s or notes by two of his teachers indicating their approval of some astrolabes that he made. The two teachers, Ibrāhīm ibn Mamdūd al-Ḥāsib and Ḥasan ibn ʿAlī al-Fihrī, wrote enthusiastically about the various astrolabes made by al-Ashraf and described each one in considerable detail. From their remarks it seems that al-Ashraf made at least four astrolabes. One of these, a sexpartite astrolabe made in 690 Hijra, survives in the Metropolitan Museum of Art in New York.[43]

[37] On the instruments of al-Ḥākim see *Sayili*, pp. 130-156, and my article "*Ibn Yūnus*" in *DSB*.

[38] On al-Farghānī see the article in *DSB* by A. I. Sabra. On his astrolabe tables see *King* 5, pp. 53-55. A survey of all known Islamic tables for constructing astrolabes is in preparation.

[39] On the *Īlkhānī Zīj* see *Kennedy* 1, no. 6, and *Storey*, pp. 58-60. On the transmission of a copy of this *zīj* to the Yemen see the anecdote related by Ibn Abi l-Faḍāʾil discussed in Section 4 in Part I.

[40] On al-Khwārizmī see the article by G. Toomer in *DSB*. His sundial tables are extant in MS Istanbul Aya Sofia 4830, fols. 231v-235r, copied 626H. A survey of these and all later Islamic tables for marking sundials is in preparation: see already the brief remarks in *King* 5, pp. 51-53.

[41] For al-Marrākushī's tables for Cairo see *Sédillot-père*, II, pp. 454 and 491. An analysis of these tables is contained in *Janin & King* 2, Section IV.

[42] On some other early Islamic references to the compass see *Janin & King* 1, p. 216, and the references there cited. Al-Ashraf's account is now studied in *Banerjee & Sabra*.

[43] The text of these *ijāzas* is published already in *Jaza iri* (see note 34 above). The New York astrolabe (on which see already *Gunther*, I, p. 243 and *Mayer*, pp. 83-84) is analyzed together with the *ijāzas* in *King* 8.

9. Muḥammad ibn Aḥmad, known as Abu l-ʿUqūl

Considerable obscurity surrounds the life and work of the Yemeni astronomer known by the curious name Abu l-ʿUqūl.[44] A marginal note in MS SB in the hand of the Sultan al-Afḍal (see 9.3) enables us to identify him as the author of the *Mirʾāt al-zamān*, the corpus of tables for timekeeping computed for Taiz, and to associate him with the Sultan al-Muʾayyad (*fl.* ca. 1300) (see 10). In MS LA of the *Mukhtār Zīj* (see 9.1) a single note identifies the author as Abu l-ʿUqūl, and since the tables attributed to him in MS SB are not contained in either the *Mukhtār Zīj* or the *Mirʾāt al-zamān* it appears that he compiled a second *zīj* (see 9.2).

The only reference to Abu l-ʿUqūl known to me in medieval Arabic sources other than astronomical manuscripts occurs in the geographical work *Taqwīm al-buldān* compiled by the early fourteenth-century Syrian scholar Abu l-Fidāʾ. In his lists of coordinates of localities in southern Arabia Abu l-Fidāʾ attributes the following values to Abu l-ʿUqūl:[45]

	Long.	Lat.
Taiz	64;30°	13;0°
Dimlua	64;40	14;5
Jubla	65;0	13;10

Now none of the available Yemeni astronomical sources use the parameter 13;0° for the latitude of Taiz, and no tables are known to have been prepared for the smaller towns of Dimlua and Jibla. Indeed, the two values 13;40° and 13;37° seem to have been used for Taiz by Abu l-ʿUqūl: the former underlies the tables for Taiz in MS LA of the *Mukhtār Zīj* and the second underlies a table attributed to him in MS SB of the astronomical compendium of the Sultan al-Afḍal.

9.1 al-Zīj al-Mukhtār min al-azyāj (B5)

MS LA, copied in 1599 probably in the Yemen, is a unique copy of this Yemeni *zīj*, which is of considerable interest.[46] The title is intended to indicate that the work was derived from other *zīj*es: in fact, it appears to be derived almost entirely from a *zīj* of Ibn Yūnus other than the *Ḥākimī Zīj* and it contains material due to or related to the Egyptian astronomer that is not extant in other known sources. The name of the author is not stated on the title folio or in the introduction, but is mentioned in a note on fol. 70r that states "these (planetary mean motions) were stated by Ibn Yūnus in his *Zīj* known as *al-Ḥākimī*

[44] I venture to guess that his name was intended to mean "the genius." (*Abū*, literally "father" here means "possessor of . . . ," and *ʿaql*, the singular of *ʿuqūl*, means "mind, intelligence, understanding.")

[45] *Abu l-Fidāʾ*, p. 90. The same author, who himself was a patron of the sciences, does not mention the astronomical activity of the Rasulid sultans of the Yemen in his discussion of them in his historical work, *al-Mukhtaṣar fī akhbār al-bashar*.

[46] The *Mukhtār Zīj* is listed as no. 57 in *Kennedy* 1, where it is incorrectly described as Egyptian and dated ca. 1200, following *Brockelmann*, I, p. 864 and the British Museum catalogue.

and were adopted by Abu l-ᶜUqūl in his *Zīj* known as *al-Mukhtār*."

The *Mukhtār Zīj* contains an introduction of forty-five chapters and extensive tables (200 pages in MS LA). The calendrical, planetary, and most of the spherical astronomical tables were taken from Ibn Yunūs, although this is not stated. Some of the tables have their counterparts in the *Ḥakimī Zīj*, but others appear to have been taken from another work by Ibn Yūnus.[47] The *Mukhtār Zīj* is an extremely important source for supplementing our knowledge of Ibn Yūnus' works: it contains material due to the earlier Egyptian scholar that is not preserved in the two extant fragments of the *Ḥākimī Zīj*, such as Ibn Yūnus' star catalogue and his lunar crescent visibility theory. The *Mukhtār Zīj* also contains tables of standard spherical astronomical functions, carefully computed for latitudes 13;0° (Aden), 13:40° (Taiz), 14;0° (Zabid), and 14;30° (Sanaa), using obliquity 23;35°.

9.2 Zīj Abi l-ᶜUqūl (B6)

MS SB (see 18.1 below) contains a few tables for timekeeping based on the distinctive value 13;37° for the latitude of Taiz, and other tables for astrology. In marginal notes to one of the former and to the astrological tables it is stated that they were taken from the *Zīj Abi l-ᶜUqūl*. The tables are quite different from any in the *Mukhtār Zīj* (see 9.1). Also, since in both the *Mukhtār Zīj* and the *Mirʾāt al-zamān* (see 9.3) the latitude of Taiz is taken as 13;40°, it may be that these tables are all that remains of a second *zīj* of Abu l-ᶜUqūl. (See also 8.2 and 17.1 on the distinctive parameter 13;37°.)

9.3 Kitāb Mirʾāt al-zamān (C2)

MS BN_3 (fols. 16r-166r) is an incomplete copy of a corpus of tables for timekeeping for the latitude of Taiz, bound in considerable disorder. The first folio of the introduction is missing but at the beginning of the next folio (fol. 16r) the author states that he has called his work *Mirʾāt al-zamān*. The tables in MS BN were copied ca. 1700 by Yūsuf ibn Yūsuf al-Maḥallī (see 38). MS MB, copied about 1600 (?), is another fragment of this corpus, containing some 95 folios of tables and bearing the title *Tuḥfat al-awān al-muntazaᶜa min Mirʾāt al-[zamān]*. MS SB of the astronomical compendium of the Sultan al-Afḍal (see 18.1) contains a few more tables from the corpus that are not preserved in either of MSS BN or MB. In the margin of one of these tables in MS SB is the following important note in the hand of the Sultan al-Afḍal:

الحمد لله وحده منقول من كتاب مراة الزمان تصنيف محمد بن احمد
الحاسب ابى العقول وضعه فى ايام جدى الملك المويد رحمة الله عليه

[47]The extensive lunar equation tables in the *Mukhtār Zīj*, which are related to the yet more extensive and more sophisticated tables attributed to Ibn Yūnus elsewhere, have been analyzed in *King* 3, pp. 134-135. An analysis of the Mukhtār lunar crescent visibility theory, which is due to Ibn Yūnus, is contained in a survey of medieval lunar visibility theories currently in preparation.

which translates:

> "Praise to God alone . . . (This is) taken from the Kitāb Mir'āt al-zamān "Book of the Mirror of Time," compiled by Muḥammad ibn Aḥmad al-Ḥāsib ("the calculator") Abu l-ʿUqūl in the days of my grandfather the King al-Mu'ayyad, may God's mercy be upon him."

Thus Abu l-ʿUqūl is identified as the compiler of the *Mir'āt al-zamān* and can be dated to the beginning of the fourteenth century. This means that the Taiz corpus was compiled after the Cairo corpus of tables for timekeeping and before the Damascus corpus. It is already well established that Abu l-ʿUqūl relied heavily on the works of Ibn Yūnus (see 9.1), but I find it curious that the Taiz corpus, with well over 100,000 entries, is considerably more extensive than the Cairo or Damascus corpuses, each with about 30,000 entries, and wonder whether Abu l-ʿUqūl was inspired by other timekeeping tables for other localities that have not survived in the known manuscript sources. It seems unlikely that Abu l-ʿUqūl conceived his tables himself, although there can be little doubt that he computed them himself or with a team of professional calculators. The underlying parameters are 13;40° for the latitude of Taiz and 23;35° for the obliquity. In his introduction Abu l-ʿUqūl lists the various tables contained in the corpus, and this enables us to some extent to sort out the confusion of tables in our three manuscript sources.

I have analyzed the Taiz corpus in some detail elsewhere.[48] For the present discussion I restricted myself to the following remarks. The tables consist of two main sets, one for timekeeping by day and the other for timekeeping by night, whereas the Cairo and Damascus corpuses serve mainly timekeeping by day.[48a] There are tables displaying the time since sunrise (MS BN only) and hour-angle as functions of solar altitude and solar longitude, each with about 13,000 entries in equatorial degrees and minutes for each degree of both arguments, and another with over 5,000 entries displaying the longitude of the horoscopus as a function of solar altitudes in the east and west and solar longitude (MS BN only). The tables for nighttime display the altitudes at daybreak of various prominent stars (MS BN only), with about 7,200 entries in degrees and minutes for each degree of solar longitude, and the altitudes of such stars as a function of the longitude of the horoscopus (MS BN only), with about 7,200 entries in degrres and minutes for each degree of argument. Other tables display the longitudes of the astrological houses for Taiz as a function of the longitude of the horoscopus (MS SB only).

9.4 Al-Yawāqīt fī maʿrifat al-mawāqīt (G2)

MSS LI_2, MA, and VG are copies of a simple almanac with this title attributed to Abu l-ʿUqūl. In the first two sources it is stated that this almanac is taken from the *Zīj* of Abu l-ʿUqūl, which may mean his second *zīj* (see 9.2) since there is no such almanac in the *Mukhtār Zīj* or the *Mir'āt al-zamān* (see 9.1 and 9.3).

[48] See *SATMI*, I and II. An extract from Abu l- Uqul's tables is also displayed in *King*, 5, p. 46.

[48a] On these see *King* 2 and 7, respectively.

10. Al-Sulṭān al-Mu᾿ayyad Dā᾿ūd ibn Yūsuf

Al-Mu᾿ayyad was the son of the Sultan al-Muẓaffar (see 5a) and brother of the Sultan al-Ashraf (see 8). He ruled the Yemen from 1297 to 1321 and was the patron of Abu l-ʿUqūl (see 9 above). See also the almanac described in 11 below which may be due to al-Mu᾿ayyad himself.

10.1 Risāla fī ʿAmal al-asṭurlāb (E4)

MS SB contains a short treatise on the astrolabe. In the margin of the first page of this treatise a marginal note in the hand of the Sultan al-Afḍal (see 18.1) states that the work was compiled by the Sultan al-Mu᾿ayyad. The work contains a few diagrams but no tables, and the star positions mentioned in the text are stated to have been computed for Ramaḍān, 680 Hijra (= December, 1281).

11. Anonymous (ca. 1325)

11.1 Almanac and ephemerides for Sanaa, 727 Hijra (D1) (G3)

MS TG_2 contains a Yemeni almanac for the year 727 Hijra (= 1326/27). The first page bears the name of the Rasulid Sultan al-Mu᾿ayyad Dā᾿ūd ibn Yūsuf (see 10 above), but there is no statement that it was he who compiled the almanac. The work begins with extensive astrological tables of a non-numerical kind, followed by descriptions of the lunar mansions and calendrical information for the Syrian year. These are followed in turn by four horoscopes computed for three dates in 727 Hijra and for the first day of 728 Hijra; one of the former is stated to have been computed using the *Ḥākimī Zīj*.

The ephemerides display positions of the sun, moon, planets, and lunar node, expressed in zodiacal signs, degrees and minutes for each day of the year. The hours of daylight and the meridian altitude are also shown: the parameters underlying the latter are 24° for the obliquity, as used by the Sultan al-Ashraf in his astrological treatise *al-Tabṣira* (see 8.1 above), and 14;30° for the latitude, which is probably intended to serve Sanaa. For each day of the year the astrological significance of the positions of the moon and planets is also indicated. For each month, information is given on the new moon, such as the elongation of the sun and moon and the lunar latitude at the time of first visibility.

The ephemerides are followed by diagrams giving information about the rising and setting and appearance of the moon, and qibla, and the rainbow. The almanac concludes with a computation for the lunar eclipse of March, 1327, and a horoscope for the vernal equinox the same month, computed with the *Ḥākimī Zīj*.

11a. Al-Sulṭān al-Mujāhid ʿAlī ibn Dāʾūd ibn Yūsuf

Al-Mujāhid was Sultan of the Yemen during the period 721-764/1321-1363. Brockelmann mentions a treatise by al-Mujāhid on the astrolabe.[49] This is an error: cf. notes 10 and 12 to Part I.

12. Ibn al-Mushrif[50]

12.1 Zīj for Zabid (B7)

MS SB (see 18.1 below) contains two tables from the *Kitāb Ibn al-Mushrif* and the *Zīj Ibn al-Mushrif*. The first is a star catalogue dated 725 Hijra (= 1325), which includes in addition to the ecliptic coordinates of thirty stars, the longitudes of the points which rise and set with the stars and the ḥalf arc of visibility as well as the meridian altitude for Zabid. The latitude underlying the values in this last table is about 14;0° and it is stated that these stellar coordinates underly "the altitude tables." This may mean that Ibn al-Mushrif compiled some tables of stellar altitude for Zabid, or that he lifted some of the tables of Abu l-ʿUqūl (see 9.4 above). The second table attributed to Ibn al-Mushrif displays the true solar longitude as a function of the mean solar longitude. The underlying parameters are a solar apogee longitude of exactly 90° and Ibn Yūnus' distinctive value 2;0,30° for the maximum solar equation.

13. Aḥmad ibn Muḥammad al-Azharī al-Bakhāniqī

Al-Bakhāniqī is known to have worked in Cairo, where he made an edition of the main corpus of tables for astronomical timekeeping. He also wrote several treatises on instruments.[51]

13.1 Kitāb Tatmīm ʿamal al-asṭurlāb (E5)

MS DT contains an extensive set of tables for marking the curves on the plates of astrolabes and brief instructions for their use, compiled by al-Bakhāniqī for Abū Jaʿfar ʿUmar, the *wazīr* and *qāḍī* of the Rasulid Sultan al-Mujāhid (see 11a). A fragment of these tables is

[49] *Brockelmann*, II, p. 242, and SII, p. 252 (also mentions a work on the raising of horses).

[50] The name is written in MS SB something like l-b-r m-s-r-b/f without any diacritical points, and Ibn Musrif seems the best reading. This Yemeni astronomer is not to be confused with the early fifteenth-century Egyptian astronomer Zayn al-Dīn Abū Bakr ibn Ismāʿīl ibn al-Mushrif (*Brockelmann*, SI, p. 869), whose various works are discussed in *Cairo Cat.*, Survey, no. C43; *MAES*; and *SATMI*, I and II.

[51] On his edition of the Cairo corpus see *King* 2, pp. 388-389 and the more detailed discussion in *SATMI*, I and II. Further information on al-Bakhāniqī and his works is contained in *MAES* and *Cairo Cat.*, Survey, no. C28.

in MS YB. Al-Bakhāniqī's astrolabe tables are an extension of the earlier tables of the ninth-century Baghdad astronomer al-Farghānī, serving each degree of latitude from 0° to 90°.[52] The generation of such tables from al-Farghānī's auxiliary table is trivial, and al-Bakhāniqī even reproduced al-Farghānī's table for marking stellar positions on the astrolabe rather than prepare a new one to incorporate five centuries of precessional motion.

14. Sirāj al-Dīn Abū Bakr ibn ʿAlī ibn Mūsā al-Hāmilī al-Yamanī al-Ḥanafī[53]

Died in 765/1363-64 (*HK*, II, p. 24) or 769/1367-68 (*HK*, V, p. 454). See also Appendix, no. 4 on his mathematical works.

14.1 A *Qaṣīda* on the lunar mansions (A5) by al-Hāmilī is in MS $ZR_{1,2,3}$ (three copies). Not examined.

14.2 An *Urjūza* on the zodiac in MS ZR_4. Not examined.

15. Abū Muḥammad ʿAbd Allāh ibn Asʿad ibn ʿAlī al-Yāfiʿī al-Tamīmī al-Shāfiʿī

Al-Yāfiʿī was born in 698/1298 in the Yemen and began his studies in Aden. From 718/1318 he lived in Mecca and Medina and in 724/1324 travelled to Jerusalem, Damascus, and Cairo. In 738/1338 he revisited the Yemen and lived in Mecca until his death there in 768/1367. Most of his works deal with mysticism but he also wrote on magic (see MS Escorial 946,10) and astronomy.[54]

15.1 Sirāj al-tawḥīd al-bāhij al-nūr fī tamhīd ṣāniʿ al-wujūd muqallib al-duhūr wa-maʿrifat adillat al-qibla wa-l-awqāt (A4)

MSS YO (two copies) and SR contain an extensive treatise with this title by al-Yāfiʿī, arranged in twenty sections (*faṣl*). The treatise deals with the standard topics of folk astronomy and begins with a poem (*qaṣīda*) entitled *ʿUqd al-durar* "The pearl necklace."

[52] On al-Farghānī see note 38 to Section 8 above.

[53] *Suter*, no. 260; *Brockelmann*, II, p. 236 and SII, p. 240. Brockelmann states that a *qaṣīda* in MSS Berlin Ahlwardt 7847 and 8261 is by al-Hāmilī's father.

[54] See the article "al-Yāfiʿī" by F. Krenkow in EI_1; *Brockelmann*, II, pp. 226-228 and SII, pp. 227-228; and *Sayyid*, pp. 146-147. Brockelmann lists MS SR of the *Sirāj al-tawḥīd* and several copies of the poem on the Syrian months.

15.2 Manẓūma fi l-shuhūr al-Rūmīya (A4)

MSS YJ (two copies), LI_3, SQ, and YH_5 are copies of a poem by al-Yāfiʿi on the Syrian months and the foods that one should eat during each month.

16. Anonymous

16.1 Zīj for Sanaa (undated) (B8)

In MS SG of the *Zīj al-Muthannā* (see 37.1 below) there have been bound some additional folios in a different and earlier hand (82r-87v) from a Yemeni *zīj* of considerable interest. The tables are for computing solar and lunar eclipses and for determining the astrological houses for the latitude of Sanaa. The solar eclipse tables are those of the Abbasid *Mumtaḥan Zīj*.[55] See 20 below.

17. Anonymous (ca. 1350)

17.1 Zīj for Taiz (B9)

MS PA, preserved in Paris, is a unique copy of a *zīj* of considerable interest which was compiled in Taiz about the year 1350. The title folio and beginning of the text of the *zīj* are missing. In the chapter on solar eclipses (fol. 33r) the author presents calculations for an eclipse of the end of Ramaḍān, 751 Hijra (November 30, 1350), and uses what he calls *al-Zīj al-Ḥākimī li-ṭūl Taʿizz*, "the *Hakimi Zīj* for the longitude of Taiz" to find the solar longitude. However, both the text and the tables in this *zīj* were compiled from several sources. The solar and lunar mean motion tables are stated to be based on the parameters of the *Ḥākimī Zīj* of Ibn Yūnus; the tables for Mercury are stated to to have been taken from the *Shāwī Zīj*, perhaps to be identified with the *Shāhī Zīj* of Naṣīr al-Dīn al-Ṭūsī(?), which was compiled at Maragha in Persia in the mid-thirteenth century; and the tables for the remaining four planets are stated to have been taken from the *Muntaḥal (?) Zīj*, a work otherwise unknown to the literature.[56]

MS PA also contains tables of oblique ascensions for each degree of latitude from 1° to 16° and for the latitudes of Taiz (13;37°–see 8.2 and 9.2) and Zabid (14;20°), as well as a so-called *ṭaylasān zīj*, that is, a table displaying the time since sunrise in seasonal day hours as a function of solar altitude for any latitude (based on an approximate formula).[57] The

[55] The *Mumtaḥan* solar eclipse tables have been analyzed in *Kennedy & Faris*. Some more copies of these tables in Egyptian sources are listed in *Cairo Cat.*, II, Section 2.1.2.

[56] The thirteenth-century Egyptian *Muṣṭalaḥ Zīj* (*Kennedy*, no. 47; extant in MSS Paris B.N. ar. 2513 and 2520) also contains tables from both the *Ḥākimī Zīj* and *Shāhī Zīj* (*Kennedy* 1, no. 32; see also *HK*, II, p. 969). See further *Cairo Cat., Survey*, nos. C12 and C13.

[57] Several tables of this kind are discussed in *SATMI*, I.

folios between the end of the text and the beginning of the tables have been filled in with some spurious material, which is in fact taken from the text of the *Ḥākimī Zīj*, and the fly-leaves are covered with notes in a script which looks like Armenian. This *zīj* deserves detailed study.

18. Al-Sulṭān al-Afḍal ʿAbbās Ibn ʿAlī Ibn Dāʾūd

A son of the Sultan al-Mujāhid (see 11a), he died in 778/1376.[58]

18.1 Astronomical compendium (B10, C3, D2, E6, G4)

MS SB, copied ca. 777/1376,[59] is an encyclopaedic work of great interest compiled by the Sultan al-Afḍal. The manuscript contains extensive sections on astronomy, mostly culled from earlier sources. It is copied in an elegant hand and bears marginalia in the hand of al-Afḍal himself. A detailed description of this astronomical material folio by folio would be desirable. For the present a brief summary must suffice.

There are numerous passages and tables taken from the *Kitāb al-Mabādiʾ wa-l-ghāyāt* of the late thirteenth-century Cairo astronomer Abū ʿAlī al-Marrākushī (see 8.2), and various tables taken from the thirteenth-century Egyptian *Muṣṭalaḥ Zīj*,[60] as well as tables from the Taiz corpus of Abu l-ʿUqūl and from his second *zīj* (see 9.3 and 9.2). The manuscript also contains some geographical tables by the late twelfth-century Syrian astronomer Ibn al-Dahhān,[61] who is known to have compiled a *zīj* now no longer extant, and some tables attributed to an unknown Yemeni astronomer named Ibn al-Mushrif (see 12), and some planetary tables taken from a work entitled *Kitāb al-Shams al-Harīrī* (?). Other tables are taken from the *Muẓaffarī Zīj* of al-Fārisī (see 6.3) and the *Īlkhānī Zīj* (see 8.2). An almanac displaying various spherical astronomical functions for each day of the solar year is introduced in the name of al-Afdal but the same almanac occurs in the *Mirʾāt al-zamān* of Abu l-ʿUqūl (see 9.3). A short treatise of one page is also attributed to al-Afḍal himself and deals with a celestial sphere that he made in 776/1374. A table displaying the solar longitude for each day of the Persian year is stated to have been compiled by al-Afḍal in 777 Hijra. A treatise on the astrolabe is attributed to the Sultan al-Muʾayyad (see 10).

A more detailed description of the contents of this manuscript would obviously be very worthwhile, but the need for investigations of the works of al-Marrākushī, the Egyptian *Muṣṭalaḥ Zīj*, and the Yemeni *Muẓaffarī* and *Mukhtār Zīj*es, is more urgent.

[58] On al-Afḍal see *Brockelmann*, II, p. 235 and SII, p. 236 and *Sayyid*, pp. 148-150 (no scientific works mentioned).

[59] The marginalia appear to be in the hand of al-Afḍal, who died in 778/1376, and in the treatise on the sphere made by al-Afḍal and in his solar longitude table the dates 776/1374 and 777/1375 are specifically mentioned.

[60] On the *Muṣṭalaḥ Zīj* see note 56 to Section 17 above. On the double-argument lunar equation tables in MS SB see *King* 3, pp. 132 and 135.

[61] On Ibn al-Dahhān see *Suter*, no. 310; *Kennedy* 1, no. 89; and *MAES*. His tables in MS SB are discussed in greater detail in *SATMI*, II.

18a. Al-Sulṭān al-Ashraf Ismāʿīl ibn al-ʿAbbās[62]

A grandson of the Sultan al-Mujāhid (see 11a) and son of the Sultan al-Afḍal (see 18), born 761/1359, died 803/1400. A. Sayyid records that astronomy was one of his interests. See also 19 below, and Appendix, no. 5.

19. Abū Bakr ibn Abi l-Maʿālī[63]

19.1 Mudkhal al-taʿlīm fī inshā᾽ al-tasyīr wa-ʿamal al-taqwīm (D3)

MS YI, copied in Istanbul in 1741, contains an astronomical poem compiled by Ibn Abi l-Maʿālī for the Rasulid Sultan al-Ashraf Ismāʿīl (see 18) in 794 Hijra (= 1391). MS TL is a later anonymous copy of the same work, also not of Yemeni provenance. The poem deals with the standard topics of medieval planetary astronomy, including the determination of planetary positions, conjunctions, lunar visibility, and eclipses.

20. Al-Kaʿbī

20.1 Zīj (B11)

In MS ZK of the *Zīj* of al-Daylamī (see 28.1) there is a single table of which it is stated that it is taken from *Zīj al-Kaʿbī*. The table displays the distance traversed by the sun in one day as a function of the solar longitude. Al-Kaʿbī's *Zīj* is not known from other sources, but may be identical to the *zīj* listed in 16.1 above.

21. Anonymous (ca. 1400)

21.1 Al-Zīj al-Mumtaḥan al-ʿarabī (B12)

MS CZ_1 contains an anonymous recension of the *Muẓaffarī Zīj* (see 6.3 above) with all of the planetary mean motion tables converted to the Hijra calendar, which, the author states, was more widely used in the Yemen than the Persian calendar. The work is entitled *al-Zīj al-Mumtaḥan al-ʿarabī*, the term *ʿarabī* referring to the use of the Hijra calendar. The introduction consists of thirty chapters and the tables other than those for the planetary mean motions are taken from the *Muẓaffarī Zīj*. The work appears to have been compiled about the year 800 Hijra (ca. 1400), since the planetary apogee positions given in the titles of the tables for the planetary equations are stated to be for 801 Hijra.

[62] *Brockelmann*, SII, p. 237 and p. 253; *Sayyid*, pp. 157-158 (no scientific works mentioned).
[63] Listed in *Brockelmann*, SII, p. 253.

22. Anonymous (ca. 1405)

22.1 Almanac and ephemerides for Taiz, 808 Hijra (D3 and G5)

MS TC contains a Yemeni almanac for the year 808 Hijra (= 1405-06), which was during the reign of the Sultan al-Nāṣir Ahmad. The work begins with almost thirty pages of lists of important events in Yemeni history and corresponding dates.[64] These are followed by various astrological tables of a non-numerical kind and a series of horoscopes for the vernal equinox, the four seasons, and each Muslim month, as well as the 19th (solar) birthday of the Sultan and the lunar eclipse of June, 1406. The ephemerides are of the same kind as the earlier set in MS TG_2 (see 11.1 above). The length of daylight, solar meridian altitude, and solar altitude at the afternoon prayer, are also displayed for each day of the year; the underlying parameters are obliquity 23;35° and latitude 13;40°, as used for Taiz in the *Mukhtār Zīj* of Abu l-ᶜUqūl (see 9.1 above). The almanac concludes with a list of the astrologically significant configurations of the planets (*taysīrat*) and the influence these are expected to have on the affairs of men (*taʾthīrāt*), for each day of the year.

23. Ismāᶜīl ibn ᶜAṭīya al-Najrānī

Al-Najrānī is reported by the seventeenth-century Yemeni biographer Ibn Abī Rijāl to have died in 794 Hijra (= 1391/92).[65]

23.1 Zīj al-Najrānī (B13)

Al-Najrānī's *zīj* has not survived in its entirety. Fragments containing only planetary mean motion tables and equation tables are preserved in MSS DA_2 (see 35.1) and MV_6,[65] and al-Daylami also incorporated some of al-Najranī's tables in his *zīj* (see 28.1). Al-Najrānī appears to have relied rather heavily on Ibn Yūnus.

24. Zayd ibn ᶜAṭīya al-Najrānī

It appears that an individual named Zayd ibn ᶜAṭīya al-Najrānī, probably a brother of Ismāᶜīl, also dabbled in astronomy. A passage from MS MV_6 (see 23.1) recorded by Griffini[66] is worth translating here:

> "Zayd ibn ᶜAṭīya said: I checked the maximum half excess of daylight for the latitude of Sanaa with these four chords and obtained 6;29, which is as in the table. I also checked it using the Indian value of the obliquity and obtained 6;36,44."

[64] The importance of these lists is noted in *Sayyid*, pp. 159-160.

[65] See further Griffini's notes on MS MV_6 in *RSO*, 7 (1916-18), pp. 123-127.

[66] *RSO*, 7 (1916-18). p. 127. The passage ends with the words *iᶜlam dhālika manqūl Battānī*, "know that this is taken from al-Battānī," which refers to other parts of MS MV_6 rather than this quotation.

The writer is clearly using the formula[67]

$$\sin d = \frac{\sin\varphi}{\cos\varphi}\cdot\frac{\sin\epsilon}{\cos\epsilon}(= \tan\varphi\ \tan\epsilon)$$

to compute the maximum half excess of daylight d from the latitude φ and obliquity ϵ. Using φ = 14;30° for Sanaa and al-Battani's parameter ϵ = 23;35° I compute d = 6;29°, which is the value given in al-Battānī's table of $d(\varphi)$ (*Nallino*, II, p. 59). Using φ = 14;30° and the Indian parameter ϵ = 24° I compute $d \approx$ 6;37°. If instead of messing about with numbers Zayd ibn ʿAṭīya had taken the trouble to measure the latitude of Sanaa he would have no doubt been surprised to find that it is closer to 15½°.

24a. Al-Mahdī Aḥmad ibn Yaḥyā

Died 840/1437 (Ahlwardt).

24a.1 A poem on the lunar mansions is in MS BW_1 (not inspected).

25. Al-Hādī ila l-Ḥaqq ʿIzz al-Dīn ibn al-Ḥusayn ibn al-Muʾayyad

A Zaydi *imām*, born 845/1441, died 900/1494.[68]

25.1 MSS YW and YX, neither of which I have examined, contain a *qaṣīda* on the lunar mansions by al-Hādī, in the first source with title *al-Sharīda ilā dhikr shuhūr al-Rūm*.

26. ʿAbd Allāh ibn ʿUmar ibn ʿAbd Allāh ibn Aḥmad, Bā Makhrama[69]

26.1 Kitāb al-Shāmil fī dalāʾil al-qibla wa-ḥisāb al-Rūm wa-l-manāzil (A6)

MSS YN, copied ca. 1700, and TH_2, copied in 1320/1902-03, are two Egyptian copies of the same Yemeni treatise on folk astronomy. In the first manuscript, which is defective, the work is attributed to Abū Hamad Maḥfūz ibn ʿAbd al-Raḥmān al-Ḥaḍramī. In the second the author is named as (*al-shaykh al-jalīl*) Bā Makhrama, and the colophon states that the work was completed in 907 Hijra (= 1501/02). The work is arranged in seventeen *faṣl*s and in the introduction the author states that his book is a summary of the writings of al-Yāfiʿī (see 15) "and others."

[67] Cf., for example, *Nallino*, I, p. 180, or *King* 1, pp. 127-128, on this formula as used by al-Battani and Ibn Yunus.

[68] See *Brockelmann*, II, p. 240, and SII, p. 248, and the references there cited.

[69] *Brockelmann*, SII, pp. 239-240, mentions the members of a Bā Makhrama family who lived in Aden *ca.* 1500.

26.2 Jadwal fī maʿrifat ittifāq al-maṭāliʿ wa-khtilāfihā fī ruʾyat al-ahilla

MS YL (one folio) contains a very simple non-numerical table attributed to ʿAfīf al-Dīn ʿAbd Allāh ibn ʿUmar ibn ʿAbd Allāh ibn Aḥmād (Bā) Makhrama purporting to display the localities where (the latitudes and hence) the oblique ascensions are sufficiently similar that if the lunar crescent is sighted in one it will be sighted in the other. Thus, for example, Mecca and Medina are listed as places where the ascensions are similar (*al-mawāḍiʿ al-muttafiqa*), and Mecca and Aden are listed as places where the ascensions differ considerably (*al-mawāḍiʿ al-mukhtalifa*).

26.3 Fawāʾid fī maʿrifat al-aẓlāl li-ʿarḍ ʿAdan wa-li-ʿarḍ Taʿizz

MS TH_7 contains some notes on the midday shadow lengths at Aden and Taiz attributed to ʿAfīf al-Dīn ʿAbd Allāh ibn ʿUmar Bā Makhrama. Values for Aden, stated to be to base $6\frac{1}{2}$, are given for each ten days of the Syrian year, and vary between $4\frac{5}{6}$ units on Kānūn I 10 and $1\frac{1}{5}$ units (south) on Ḥazīrān 10. The values for Taiz are given for two months only, and Bā Makhrama concludes his discussion saying *wa-ʿalā hādhā fa-qis al-bāqī*, "measure the remaining (shadows) in the same way."

27. Nūr al-Dīn ʿAlī ibn ʿAbd Allāh al-Ṭawāshī (?)

Not mentioned in the modern sources.

27.1 Miftāḥ al-asrār fī ʿilm al-falak al-dawwār (A7)

MSS YR and YT are copies of a treatise with this title arranged in 10 *bāb*s dealing with folk astronomy and simple timekeeping.

28. Muḥammad ibn ʿAlī ibn Muḥammad al-Daylamī

A son of the Imām al-Wāthiq bi-llāhi, who was a descendant of the eleventh-century Imām al-Nāṣir Abu l-Fatḥ al-Daylamī.[70] Muhammad al-Daylamī's *Zīj* was apparently compiled about the year 1520.

[70] The full name of the author of the *Zād al-musāfir* is given in MS ZK_1 as ʿIzz al-Dīn Muḥammad ibn al-Imām al-Wāthiq bi-llāhi ʿAlī ibn Muḥammad ibn ʿAlī ibn ʿAbd Allāh ibn Ibrāhīm ibn Sulaymān ibn Mūsā ibn Muḥammad ibn Sharaf al-Dīn al-Ḥasan ibn Amīr al-Muʾminīn al-Imām al-Nāṣir li-Dīn-Illāh Abi l-Fatḥ al-Daylamī. *Azzawi*, p. 338 mentions the title without indicating his source.

28.1 al-Zīj al-mukhtaṣar fī taqwīm al-kawākib al-khams(a) wa-l-shams wa-l-qamar = Zād al-musāfir (B14)

MS ZK_1 copied in Ibb in 1091 Hijra (= 1680) and preserved in Zabid, contains a *zīj* by Muḥammad al-Daylamī bearing these titles. The introduction consists of ten chapters, and the extensive tables that follow are an interesting mixture. Some of them are taken from Ibn Yūnus, and others, such as the planetary equation tables, are attributed to Ibn Isḥāq al-Maghribī. The latter individual worked in Tunis in the early thirteenth century, and his *Zīj*, which was widely used in medieval Tunis, has only recently been recovered in the manuscript sources.[71] Other tables in al-Daylamī's *Zīj* are stated to be taken from the *Taysīr al-maṭālib* and *Mukhtār Zīj* (see 7.1 and 9.1; al-Daylamī does not mention the authors of these works), as well as from the *Zīj* of al-Najrānī (see 23.1). The second chapter of the *Zād al-musāfir* contains a computation of the solar longitude for a date in the year 927 Hijra (= 1521). This manuscript deserves further study. (See also 29.1 and 30.1 below.)

29. Anonymous (planetary tables)

29.1 Planetary tables for Sanaa (B15)

MS ZK_3 contains a set of planetary tables, some of which are based on the parameters of Ibn Yūnus. The planetary equation tables are unusual: the functions tabulated represent the equation of center, and the epicyclic equation at maximum distance with a correction for minimum distance. These planetary equation tables thus bear some resemblance to the Ptolemaic auxiliary lunar equation tables. They are preceded by a computation taken from the *Zīj* of al-Daylamī (see 28.1).

MS HE_4 contains a set of tables of the astrological houses for the latitude of Sanaa, also taken from an unidentified source.

30. Anonymous (astrology)

30.1 List of horoscopes of Yemeni rulers (F6)

MS ZK_2 contains a set of horoscopes of various medieval Yemeni rulers, which I have not been able to examine properly.

[71] On Ibn Isḥāq al-Maghribī see *Suter*, no. 356. His *Zīj*, which is not listed in *Kennedy* 1, is extant in MS Hyderabad State Central Library 298 (ca. 400 pp., ca. 800 Hijra), and is an extremely important source for the future study of the history of Islamic astronomy.

31. ʿAbd Allāh ibn Ṣalāḥ Dāʿir

Perhaps to be identified with the seventeenth-century Yemeni historian having the same unusual name.[72]

31.1 Tables for timekeeping by the stars (C4)

MS BN of the corpus of tables for Taiz (see C4) contains some additional tables of a less sophisticated type for the latitude of Sanaa (fols. 121v, 124r-127r). The tables are all for timekeeping by the stars (see below).[73] In the heading of the first one it is stated that the stellar coordinates were based on the observations of Ibn Yūnus, but the name of the compiler and the date for which he computed his stellar coordinates have been deliberately erased. However, it is just possible to read the name ʿAbd Allāh ibn Ṣalāḥ Dāʿir, which also appears on fol. 121r in reference to some instructions on the use of the third table.

The quantities displayed in these tables are (a) the longitude of the horoscopus as a function of the altitude of sixteen stars; (b) the longitude of the horoscopus as a function of each hour of visibility of eighteen stars; and (c) the altitudes of eighteen stars at each hour of visibility. The hours of visibility represent twelfths of the arc of visibility for a given star and hence differ for each star, and I know of no other Islamic tables based on such units of time.

32. Anonymous (folk astronomy)

32.1 Kitāb al-Iḍāḥ al-shāfī bi-l-itqān fī maʿrifat al-manāzil wa-l-azmān (A8)

MSS YV (16 pp., copied ca. 1000 Hijra) and LI_4 contain an anonymous treatise on folk astronomy with this title. The author mentions the *Qāmūs* of Majd al-Dīn (al-Fīrūzābādī),[73a] who spent the last years of his life in the Yemen, and also the *Kitāb al-Yawāqīt* by al-Janadī (see 5).

33. Muḥammad ibn ʿAbd al-Laṭīf al-Thābitī

According to the title folio in MS VB_2 al-Thābiti was a Syrian who lived in Zabid. In the introduction to his tables he states that they were compiled in the year 1047 Hijra (= 1637/38).

[72] On the historian see *Brockelmann*, II, p. 528, and SII, p. 635, and *Sayyid*, p. 222.
[73] On the tables see further *SATMI*, I and II.
[73a] On al-Fīrūzābādī see the article in EI_2 by H. Fleisch.

33.1 Prayer-tables for the Yemen (C5)

MSS BN_2, SK_1, and VB_2 contain a simple set of prayer tables compiled for the latitude of the Yemen by al-Thābitī.[74] MS AK appears to be a fourth copy of al-Thābitī's tables.

For each day of the Syrian year al-Thābitī tabulates the solar longitude and which thirteenth part of the lunar mansions is (a) culminating at sunset, (b) rising at daybreak, (c) culminating at midnight, and (d) culminating at daybreak. He also tabulates the lengths of day and night and of the midday and afternoon shadows (to base $6\frac{1}{2}$). Underlying the shadow lengths is a crude linear zigzag scheme. A rearrangement of al-Thābitī's column of entries for the lunar mansion rising at daybreak is found in an almanac called *Ḥisāb al-Shibāmī* which is still in use in the Hadramawt (see 43.1 below).

34. Aḥmad ibn Yaḥyā ibn Aḥmad al-Ṣaᶜdī al-Dawwārī

Died 1061/1651.[75]

34.1 His treatise *al-Maqṣad al-ḥasan . . .* , dealing with inheritance, tradition, and the history of the Zaydīya, apparently also treats of astronomy. The work is extant in several manuscripts, none of which I have seen.

35. Anonymous

35.1 Al-Zīj al-majmūᶜ[76] (B16)

MS DA_1 contains an anonymous *zīj* gathered from earlier Yemeni *zīj*es, mainly the *Muẓaffarī Zīj* (see 6.3). The manuscript was copied by Yūsuf al-Maḥallī (see 38), who may have been the compiler.

36. Al-Ḥasan ibn ᶜAbd Allāh al-Sarḥī

Al-Ḥasan al-Sarḥī completed his *Zīj* in 1070 Hijra (= 1659/60). Only one copy of this *zīj* is known to me, and the work clearly did not enjoy the popularity of the *Zīj* of al-Sarḥī's brother ᶜAbd Allāh (see 37 below).

36.1 Bughyat al-ṭālib al-mustafīd wa-mughni l-ḥāsib al-mufīd = Zīj al-Sarḥī (I) (B17)

MS SE is a unique copy of al-Sarḥī's *zīj*. The introduction is divided into thirty chapters,

[74] More information on these tables is contained in *SATMI*, II and III.
[75] *Brockelmann*, SII, p. 559; *Sayyid*, pp. 234-235. See also *RSO*, VII, p. 575.
[76] Not listed in *Kennedy* 1.

and virtually all of the tables other than the mean motion tables are taken from the *Muẓaffarī Zīj* (see 6.3). Al-Sarḥī mentions the works *Zād al-musāfir* (see 28.1) and *Tuḥfat al-muḥāḍir* (another Yemeni *zīj*??) in his introduction, and some of his tables were taken from the *Zād al-musāfir* rather than directly from the *Muẓaffarī Zīj*.

37. ʿAbd Allāh ibn ʿAbd Allāh al-Sarḥī, known as al-Muthannā

A brother of al-Ḥasan al-Sarḥī (see 36), who compiled a *zīj* in 1081 Hijra (= 1670). He was known as al-Muthannā because his name was the same as his father's, and he died in 1097 Hijra (= 1686).[77]

37.1 Ghāyat itqān al-ḥarakāt li-l-sabʿat al-kawākib al-sayyāra = Zīj al-Muthannā = Zīj al-Sarḥī (II) (B18)

MSS HD, HE, LE, MT (four copies), SC, SD, SG, SH, VE, and YP are copies of a *zīj* which became very popular in the Yemen. The introduction to the *zīj* is divided into forty chapters and the tables differ little from those in the earlier *zīj* of his brother.

38. Yūsuf ibn Yūsuf al-Maḥallī[78]

Considerable confusion surrounds this individual, not least because there was a contemporary Egyptian astronomer with the same name, who was nicknamed Yūsuf Kalārjī. It seems to me that we are dealing with the same individual, since the Egyptian Kalārjī did spend time in Mecca. The appellation al-Maḥallī refers to the town of al-Maḥalla in the Nile Delta.

Yūsuf Kalārjī is known to us by three works apparently compiled in Egypt: an extensive treatise on sundial construction completed in 1127 Hijra, a sophisticated treatise on the lunar mansions completed in 1133 Hijra, and a set of prayer-tables for Crete.[79] In 1113 Hijra Yūsuf assisted his teacher, the Egyptian astronomer Riḍwān Efendī, to make a celestial globe, now in the Public Library, Leningrad.[80]

MS AL of the *Zīj* of al-Kawāshī (see 7.1) was copied by Yūsuf ibn Yūsuf al-Makkī al-Maḥallī al-Shāfīʿi al-Ashʿarī in Sanaa in 1143 Hijra. MS BN_1 of al-Jaḥḥāf's almanac (see 39.1) and BN_2 of Abu l-ʿUqūl's *Mirʾāt al-zamān* (see 9.3), and MS DA of the anonymous *al-Zīj al-majmūʿ* (see 35.1) are also in the same hand. MS MH of the Maghribi (?) version of

[77] The *Zīj al-Muthannā* is listed in *Brockelmann*, II, p. 537 and SII, p. 567 (where MSS LE, MT [four copies], VE, and YP are mentioned) and in *Kennedy* 1, no. X212. I am grateful to Qāḍī Alī al-Sharafī of Sanaa for kindly allowing me to photocopy his manuscript of the *Zīj al-Muthannā*.

[78] *Dorn*, pp. 34-35 and *Ahlwardt*, no. 5749, Sections 33 and 34 have Yūsuf ibn ʿAbd Allāh (d. 1153 Hijra).

[79] For details see *Cairo Cat.*, II, Sections 4/7/25; 1/4/17; 3/1/31; and *Survey*, no. D61.

[80] *Mayer*, pp. 81-82.

of the perpetual almanac of Zacuto[81] bears a mark of ownership by Yūsuf ibn Yūsuf al-Maḥallī dated 1157 Hijra. This manuscript may have been copied in the Yemen; the colophon, written in a different hand from the rest of the manuscript, and stating that the manuscript was copied by Muḥammad ibn Aḥmad Āghā Arḍrūmī in 1086 Hijra, is perhaps not to be trusted.

The following works were prepared by Yūsuf ibn Yūsuf al-Maḥallī in the Yemen.

38.1 Taqwīm al-sana (ephemerides for Sanaa for the solar year 1146-47 Hijra) (D5 and G6)

MS VD is an anonymous almanac copied by al-Maḥallī for the solar year 1146-47 Hijra (= 1733-34) dedicated to the Yemeni Sultan al-Manṣūr Abu l-ʿAbbas Ḥusayn. MS MU, which I have not examined, is a copy of the same work introduced in the name of al-Maḥallī and also copied by him. The work begins with astrological prognostications for the year, both for the world in general and Sanaa and the Yemen in particular, and continues with prognostications for various astronomical events for that year, including the equinoxes and solstices, the partial solar eclipse, the first visibility of Sirius, and some planetary conjunctions, as well as the fortieth birthday of the Sultan. Information is also given on the size, direction, and setting time of the lunar crescents for each Muslim month. The remainder of the work consists of tables displaying the positions in signs, degrees and minutes of the sun, moon, and planets for each day of the year. Additional tables indicate when relative positions of the moon and planets are astrologically significant. The compiler of the tables states that he used "the new observations" (*al-raṣad al-jadīd*): this probably means that he computed the ephemerides using the *Zīj* of Ulugh Beg of Samarqand.[82] Although no recension of this work is known to have been prepared in the Yemen, it should be borne in mind that al-Maḥallī was an Egyptian and the *Zīj* of Ulugh Beg had been used in Egypt for over two centuries.

38.2 Edition of the Kitāb Mirʾāt al-zamān

The tables of time since sunrise and hour-angle in MS BN.128v-166r of the Taiz corpus (see 9.3 above) were copied separately with values of the two functions displayed side by side. A note in MS BN.128r informs us that al-Maḥallī arranged the tables that follow in this way and copied them himself.

39. Ḥusayn ibn Zayd ibn ʿAlī ibn Jaḥḥāf[83] ? = ? Ṣārim al-Dīn Ibrāhīm ibn Zayd ibn ʿAlī ibn Ibrāhīm ibn al-Mahdī Aḥmad ibn Yaḥyā ibn Qāsim known as Jaḥḥāf[84]

There are problems with the proper name and identity of this individual. See also Appendix A, no. 9.

[81] On this manuscript see note 16 to Part 1.
[82] On this *Zīj* see *Kennedy* 1, no. 12.
[83] Cf. *Brockelmann*, SII, p. 567, where the author's dates are given as 991/1583-1065/1155.
[84] Cf. *Sayyid*, pp. 256-257, where the author's dates are given as 1075/1664-1116/1704.

39.1 Kitāb al-Yawāqīt fī maʿrifat al-mawāqīt (G7)

MSS BN_1, LZ_3, YH_3 are copies of simple almanac with this title attributed to Ibn Jaḥḥāf.[85]

39.2 Calendrical tables (H1)

MS VB_5 contains some simple calendrical tables for the period 1197-1300 Hijra attributed to Ibn Jaḥḥāf. (See further 41.1 below.)

40. Anonymous (prayer-tables)

40.1 Anonymous prayer-tables (C6)

MS SI contains a set of prayer tables computed for latitude 15° (Sanaa), displaying the times of prayer in hours and minutes for each day of the solar year, according to the Ottoman convention that sunset is 12 o'clock.[85a] Such tables were common in various parts of the Ottoman Empire, and tables for a given Muslim year were compiled annually. Thus, for example, MS LJ_5 contains a set of Yemeni prayer-tables displaying the prayer-times for each day of the year 1293 Hijra (= 1876).

Other simple Yemeni prayer-tables have been located in the following sources. MS LH_2 contains some incomplete prayer-tables, displaying what is probably the ascendant at daybreak, and also the shadows at midday and the *ʿaṣr* (to one digit for each day of the year). MS LF_2 contains some tables displaying the dates of the sun's entry into each of the zodical signs, the shadows at midday and the *ʿaṣr* at ten day intervals and the lunar mansions culminating at sunset, midnight, and daybreak, and rising at sunset, for each ten days of the year. MS LJ_4 contains some incomplete prayer-tables displaying the shadows at midday and the *ʿaṣr*, as well as the number of hours of daylight and night, to two sexagesimal digits for each thirteen days of the year.

41. Miscellaneous

41.1 Calendrical tables (H1)

MSS LF, LG, LH, LI, LJ, MR (ten sets), SJ, SK, VB, and VF all contain simple calendrical tables dating mainly from the eighteenth and nineteenth centuries. The purpose of these tables was to convert dates in the Muslim lunar calendar to the Syrian solar calendar on which most Yemeni agricultural almanacs are based. Also the times of prayer, regulated by shadow lengths and twilight, are tied to the solar calendar. (The reader should bear in mind that most early *zīj*es, including, for example, the Yemeni *Muẓaffarī* and *Mukhtār*

[85] MS LA_3 is listed in *Brockelmann*, SII, p. 567. "Cet ouvrage est beaucoup plus intéressant qu'on ne le croirait" (Landberg).

[85a] On this convention and on Ottoman tables for timekeeping see *King* 8.

Zījes (see Sections 6.3 and 9.1), contain extensive and rather sophisticated tables for converting between the Hijra, Syrian, Persian, and Coptic calendars.)

I have not investigated these late tables systematically. The following notes list authors and titles where these are available, the range of Hijra dates served by the tables, and the manuscript sigla. I have noticed that most of the tables give the Syrian date and *nota* (day of the week) corresponding to the *first day* of each month of the Hijra year, and others, give the Syrian dates corresponding to *each day* of the Hijra year. Unfortunately, I have not examined MS MR(6), which contains some calendrical tables attributed to the Sultan al-Manṣūr bi-llāh al-Qāsim ibn Muḥammad, dating from 1029 Hijra (= 1620). The earliest of the tables that I have examined is no. 2 in the list below. The manuscript was copied in Bayt al-Faqīh in 1101 Hijra (= 1689-70), and the anonymous author, when introducing his own tables for 1101-1200 Hijra, refers to earlier tables covering the period 1001-1100 Hijra that were prepared by al-Ṣiddīq ibn Muḥammad al-Khaṣṣ. The author of the tables no. 10, Jamāl al-Dīn al-Akwaᶜ, who prepared tables for the period 1201-1300 Hijra, states that he was following in the tradition of his precedessors al-Khaṣṣ, whose tables covered the eleventh Hijra century, and Ḥusayn ibn Zayd Jaḥḥāf (see also 39.2), whose tables covered the twelfth century. However, the tables of Jaḥḥāf that I have inspected, no. 6 below, cover virtually the same period as al-Akwaᶜ's tables. The author of no. 12, Muḥammad ibn Aḥmad ibn al-Imām, states that he was dissatisfied with the tables of al-Khaṣṣ and al-ᶜIlfī (no. 5) and based his own tables on the *Zīj* of al-Muthannā al-Sarḥī (see 37.1).

1. Saᶜīd ibn ᶜUmar Bā Bashīr al-Shdy (?)

   1101-1150 Hijra - MS MF

   The author states that these tables are a continuation of those prepared by his father ᶜUmar ibn ᶜAbd Allāh Bā Bashīr, approved by Muḥammad al- ᶜAydarus ibn ᶜAbd Allāh ibn Shaykh al- ᶜAydarus.

2. Anonymous

   1101-1200 Hijra - MS VF_3

3. Muḥammad al- ᶜAṭṭās ibn Salām Bā Faḍl al-Haḍramī al-Shāfiᶜī

   1128-1203/5 (?) Hijra - MS ME

   The author mentions ᶜAli ibn ᶜAbd Allāh al- ᶜAydarus and Muḥammad al-Ḍaᶜīf al-Saqqāf in his introduction.

4. Yaḥyā ibn Muḥsin ibn Aḥmad ibn Rājib

   1160-1253 Hijra: MS YH_1 (entitled *Mufīdat al-sāʾil ᶜan ḥulūl al-shams fi l-manāzil*)

5. Ibrāhīm ibn Yaḥyā al-Qurashī al-ᶜIlfī

   1160-1253 Hijra - MS MR(9) (entitled *Mufīdat al-sāʾil ᶜan ḥulul al-shams fi l-manāzil*)

6. Ḥusayn ibn Zayd Jaḥḥāf (see 39.2)

   1197-1300 Hijra - MS VB_5

7. Jamāl al-Dīn ʿAlī ibn Muḥammad ibn Ḥusayn al-Murtaḍā

   1177-1200 Hijra - MS VB_1

8. Anonymous

   1181-1253 Hijra - MSS MR(2) and LJ_1
   (also gives details of the midday and afternoon shadows)

9. Anonymous

   1183-1253 Hijra - MS VB_3

10. Jamāl al-Dīn ʿAlī ibn al-Ḥasan ibn Muḥammad al-Akwaʿ

    1201-1300 Hijra - MSS LF and LG (see also MS VF_1)

11. ʿAfīf al-Dīn ʿAbd Allāh ibn Aḥmad ibn Aḥmad al-Khayrī al-Shatmakhī al-Zabīdī

    1201-1325 Hijra - MS VB_4

12. Muḥammad ibn Aḥmad ibn al-Imām

    1215-1230 Hijra - MS LH_1

    1215-1241 Hijra - MS YS and BW_3 (both entitled *al-Nafḥa al-nadīya* and stated to be taken from the *Ghāya* of al-Sarḥī (see 37.1)

13. Muḥammad ibn Aḥmad al-Najjār al-Anṣārī

    1237-1262 Hijra - MS BW_2

14. Fakhr al-Islām ʿAbd Allāh ibn Ḥamza (died 1269/1852, see *Brockelmann*, SII, p. 817): *Bulghat al-muqtāt fī maʿrifat al-awqāt*

    1263-1300 Hijra - MS LI_1

    1257-1301 Hijra - MS LJ_3

    Also MSS MR(1,7,10) and SK_2

15. ʿAbd al-Wahhāb ibn ʿAlī al-Wārith
    Dates not recorded - MS SJ_2

16. ʿAbd al-Wāsiʿ ibn Yaḥyā al-Wāsiʿī (see 47)

    Zahr al-zuhūr

    Dates not recorded - MS SJ_1

17. Anonymous

Dates not recorded - MS BY

42. Abu l-Qāsim al-Ḥyhy (??) al-Makkī

42.1 Notes on the seasons, the prayer times, the lunar mansions, and the qibla attributed to this individual are contained in MS TH_3. The author quotes al-Yāfiʿī (see 15) and al-Fārisī (see 6) and the eleventh-century religious scholar al-Ghazālī, and in the margin of the section on the qibla there is a quotation from *al-ʿbʾb* (?) of Shihāb al-Dīn ibn (?) Aḥmad ibn ʿUmar al-Mrsd (?) al-Shāfiʿī al-Madhjihī.

43. Bā Ṣabrayn al-Shibāmī

43.1 Ḥisab al-Shibāmī (G8)

This table, published by R. B. Serjeant, displays the lunar mansions rising at dawn for each day of the Gregorian year.[86] The same table in different format is contained in the earlier prayer-tables of al-Thābitī (see 33.1).

44. Muḥammad Ḥaydara

Scholar and astronomer (*al-falakī*) at Taiz.

44.1 Taqwīm Ṭawāli al-Yaman li-sanat 1365 Hijrīya (printed in Aden)

The text of this almanac for the year 1365 Hijra (= 1945-46) has been translated by R. B. Serjeant.[87] The information given for each day of the year is mainly agricultural and meteorological. Ḥaydara states that his almanac is based on the almanac of the Imām al-Mahdī (?) and on the reckoning (*ḥisāb*) of ʿAlī ibn ʿAlī al-Yamanī, formerly *shaykh al-Islām* in Sanaa.

[86] *Serjeant* 1, pp. 434-435. The table displays the lunar mansion rising at daybreak, not the lunar mansion corresponding the position of the sun: this explains the problem noted in *Serjeant* 1, p. 436.

[87] *Serjeant* 1, pp. 444-458.

45. Al-Shawqānī

45.1 Jawāb al-sāʾil ʿan tafṣīl al-qamar manāzil

A work with this title is mentioned in al-Shawqānī's *al-Badr al-tāli* .[88]

46. Al- ʿAnsī

A device for converting dates in the Hijra and Western calendars (H2) that is currently available in the Yemen is associated with an individual named al- ʿAnsī. The device consists of a graduated circular scale that can rotate on another scale printed on a board measuring about 3 feet by 2 feet.

47. ʿAbd al-Wāsiʿ ibn Yaḥyā[89]

47.1 Zahr al-zuhūr

MS SJ_1 contains some calendrical tables[90] with this title attributed to al-Wāsiʿī (see 41.1 above).

47.2 Kanz al-thiqāt fī ʿilm al-mīqāt (C7 and H3)

The work was prepared by al-Wāsiʿī for the Imām Yaḥyā and was printed in Cairo in 1358 Hijra (= 1939/40). It contains two sets of tables, the first of which is entitled *jadwal al-sinīn*, "table of years," and displays the *nota* and the solar year day number corresponding to the first day of each lunar month for the period 1358-1600 Hijra (= 1939-ca. 2175). With such a table one can find very easily the day of the solar year corresponding to a given Hijra date and then use al-Wāsiʿī's prayer-tables, which follow. These display for each day of the solar year (a) the lunar mansion of the sun; (b) the ascendant at nightfall; (c) information about the rising or setting of a prominent star or constellation (*maʿālim al-zirāʿa*); (d) the solar longitude; (e) the corresponding date in the Syrian calendar; (f) the shadow lengths at midday and the afternoon prayer (to base 7); (g) the corresponding times in hours and minutes according to the Ottoman convention; (h) the times of daybreak and sunrise. These quantities are very carelessly computed, even though al-Wāsiʿī states that he worked for years to prepare his two tables and had consulted several *zīj*es. The shadow lengths are particularly inaccurate and al-Wāsiʿī even gives widely divergent midday shadows at the two equinoxes. Also, the length of morning twilight is taken as $1^h 10^m$ throughout the year. Al-Wāsiʿī's tables in a disguised form are still in use in the Yemen (see 48.1 below).

[88] *Serjeant* 1, p. 434, note 2.
[89] On al-Wāsiʿī see *Brockelmann*, SII, p. 821 (no scientific works mentioned).
[90] See also *SATMI*, II and III on these tables.

48. Anonymous

48.1 Modern prayer-tables for Sanaa (C8)

The prayer-tables currently in use in Sanaa (*Taqwīm al-ʿArabīya al-saʿīda*) give the times only in what the Yemenis call "Arab time," that is "Turkish time." The times for twilight are approximate: for example the *fajr* is 1^h 10^m before sunrise and the *ʿishā* is at 1 o'clock (one hour after sunset) throughout the year. At the equinoxes these durations of twilight correspond to parameters 17° and 15°. The time of the *ʿaṣr* is carelessly computed, the error being as much as 20^m. A correction of -5^m has been applied to the time of midday and -20^m to the time of sunrise (*shurūq*) throughout the year. The solar longitude and corresponding lunar mansion (*manzilat al-shams*) as well as simple information about a prominent star (*maʿālim al-zirāʿa*) are given for each day of the year. The calendar is given for the Gregorian year, and the corresponding Hijra and solar Hijra dates are also given. These tables are apparently based on those of al-Wāsiʿī discussed in Section 47.1.

48.2 Modern prayer-tables for Taiz

In the almanac currently in use in Taiz the six times are given in both systems (*ʿarabī* and *ifrankī*), and the parameters for morning and evening twilight appear to be 20° and 18°. Some correction appears to have been made for refraction. The time of the *midfaʿ*, the cannon shot marking 12 noon *ifrankī*, is also tabulated. The date is given according to the Hijra and Gregorian calendars as well as the Coptic and Jewish calendars.

APPENDIX A

YEMENI WORKS ON ARITHMETIC, SURVEYING, AND INHERITANCE

I have not investigated any of the manuscripts noted below, many of which have been listed by Suter, Brockelmann, or Azzawi. I have therefore not attempted to sort out the problems associated with the names of some of the authors and their works.

Several Yemeni informants told me the legend that "algebra was invented in Zabid." The only Yemeni work on algebra that has come to my attention in al-Khuzāʿī's commentary on the treatise of the celebrated ninth-century Baghdad astronomer al-Khwārizmī.

Most Yemeni mathematical manuscripts which I have encountered are copies of Yemeni works. However, MSS Berlin Ahlwardt 5962, Cairo Dār al-Kutub *majāmīʿ* 713,14 (fols. 136v-161r, 1013H) and Sanaa Grand Mosque West Library *tarājim* 36 (38 fols., ca. 1150H) are Yemeni copies of a simple treatise on arithmetic entitled *al-Tabṣira fī ʿilm al-ḥisāb* by the celebrated twelfth-century scholar Samawʾal al-Maghribī (*Suter*, no. 302), who worked in Northern Iran.

1. Abū Yaʿqūb Isḥāq ibn Yūsuf al-Ṣardafī al-Yamanī

Died 500/1106-07 according to Ḥājjī Khalīfa and MS Berlin 5977, fol. 31r (Ahlwardt).

References: *Suter*, no. 260; *Brockelmann*, I, p. 620, and SI, p. 855.

1.1. K. al-Kāfī fī mukhtaṣar al-Hindī = K. al-Ḍarb al-Hindī = Mukhtaṣar (al-ḥisāb) al-Hindī

On Indian arithmetic: MSS Berlin Ahlwardt 5960 and 5961 (three copies); Cairo Dār al-Kutub *majāmī* 703,4 and 704,2, *riyāḍa* 84,1, and Taymūr *riyāḍiyāt* 97; Ambrosiana D371,2 and F191; Princeton Mach 4789 = Yahuda 334; Vatican 1115 and 1139,8; Milan Ambrosiana CCLXXI,1; Alexandria 1030 B. A commentary by al-Hāmilī (see below) entitled *Maʿūnat al-ṭullāb . . .* : MS Berlin Ahlwardt 5977; a commentary by Muḥammad ibn ʿAbd Allāh (see below) entitled *Kifāyat al-muhtadā . . .* : MS Ambrosiana D 550; and some anonymous *ḥawāshī*: MS Cairo Dār al-Kutub *riyāḍa* 84,2.

1.2. al-Kāfī fi l-farāʾid

On arithmetic of inheritance (*HK*, I, pp. 1377-1378): MSS Sanaa Grand Mosque Library *farāʾiḍ* 38 (173 fols., ca. 750 Hijra); Berlin Ahlwardt 4688; Ambrosiana H93,2 and CCLXXI,3. Löfgren (I, p. 144) lists six other copies in the Ambrosiana. Commentary by

Abū ʿAbd Allāh Muḥammad ibn ʿAbd Allāh ibn ʿAbd al-Raḥmān ibn Salm (see below): MS Ambrosiana D550. Some anonymous notes on this work compiled in 800/1397 are preserved in MS Berlin Ahlwardt 5977, fol. 31. Several other commentaries are listed by Ḥājjī Khalīfa.

1.3 al-Hindī fī ʿilm al-farāʾiḍ

On inheritance (= 1.2?): MS Ambrosiana H93,3.

2. Abu l-Ḥasan Aḥmad ibn Muḥammad ibn Ibrāhīm al-Ashʿarī al-Yamanī al-Nassāba al-Ḥanafī (as in *HK*)

According to Ḥājjī Khalīfa he died ca. 500-600/1100-1200.

References: *Brockelmann*, SI, p. 558; *Azzawi*, p. 231 (from *HK*).

2.1 K. al-Tuffāḥa fī ʿilm al-misāḥa

A treatise on surveying (*HK*, I, p. 426): MSS Ambrosiana A29,2 (copied Sanaa, 916H; author called Shams al-Dīn Aḥmad ibn Ibrāhīm al-Ashʿarī al-Tihāmī), and CLXXXVII,3 (author called Abu l-ʿAbbās Aḥmad ibn Muḥammad ibn Ibrāhīm al-Ashʿarī); Florence 32,2 (Assemani CCLXXIV) (author called Abu l-ʿAbbās Aḥmad ibn Muḥammad ibn Ibrāhīm al-Ashʿarī); MS Istanbul Nurosmaniye 2524,2 (fols. 45v-95v, n.d., author called Shihāb al-Dīn Aḥmad ibn Muḥammad al-Yamanī); Hyderabad Asaf. I, 800_{177} (*Brockelmann*). A commentary on this work by Sharaf al-Dīn Yaḥyā ibn Taqi l-Dīn al-Ḥalabī (*fl.* ca. 1000 Hijra) is in MS Damascus Ẓāhirīya 7582. A poem based on al-Ashʿarī's treatise entitled *Nukhabat al-tuffāḥa* was compiled by ʿAbd al-Laṭīf ibn Aḥmad al-Dimashqī (*fl.* ca. 1150 Hijra) and exists in MS Gotha 1500 (*Brockelmann*); a commentary on the poem by the same writer is in MSS Cairo Dār al-Kutub *riyāḍa* 310, 646, etc.

3. Abū ʿAbd Allāh Muḥammad ibn Aḥmad ibn ʿUmar al-Khuzāʿī (Suter and Sezgin)? = ? Aḥmad ibn ʿUmar ibn Hāshim al-Khuzāʿī al-Yamanī (Azzawi)

Azzawi states that he died in 680/1281, quoting the *Hidāyat al-ʿārifīn*.

References: *Suter*, no. 367; *Azzawi*, p. 232; *Sezgin*, V, p. 240.

3.1 Sharḥ Mukhtaṣar al-Khwārizmī fi l-jabr wa-l-muqābala

A commentary on the algebra of the ninth-century Baghdad astronomer al-Khwārizmī: MSS Istanbul Şehid Ali Paşa 2706,5 (fols. 147-282). I have not inspected this work, which Sezgin (V, p. 240) supposes to be due to Abu l-ʿAbbās Aḥmad ibn Muḥammad known as Ibn al-Hāʾim, the prolific writer of works on arithmetic who flourished in Jerusalem ca.

800/1400 (*Suter*, no. 423). A fragment from al-Khuzāʿī's commentary relating to the arithmetic of inheritance survives in MS Cairo Ṭalʿat *majāmiʿ* 207 (fols. 52r-60v, 734H). Parts of an anonymous commentary including an introduction of considerable historical interest are in MS Hyderabad Salar Jung Museum mathematics 20 (fols. 2v +(?) 3r-52r, ca. 1000 Hijra). MS Istanbul Şehid Ali Paşa 2706,6 (fols. 283-288) contains a work entitled *Kitāb al-Inshā' fī ʿilm al-jabr wa-l-muqābala* attributed to Muḥammad ibn Aḥmad al-Khuzāʿī.

3.2 al-Muqaddima fi l-ḥisab li-ʿāmmat aḥdāth al-kuttāb

A treatise on arithmetic dealing mainly with fractions: MSS Cairo Dār al-Kutub K18362,1 (fols. 1-18v, 1001 Hijra); Alexandria Municipal Library B1030,6; and an unnumbered manuscript in al-Ahqaf Library in S. Yemen (17 fols., 1185H, Arab League S. Yemen microfilms no. 247); all attributed to Abū ʿAbd Allāh Muḥammad ibn Aḥmad ibn ʿUmar al-Khuzāʿī. MS Oxford I.918,2 (see *Suter*, no. 367) appears to be another copy of the same treatise. The work is also called *al-Mazīḥfīya* after al-Mazīḥfī (?), another appellation of its author.

3.3 Jawāhir al-ḥisāb

This title is mentioned by Azzawi, quoting the *Hidāyat al-ʿārifīn*.

4. Sirāj al-Dīn Abū Bakr ibn ʿAlī ibn Mūsā al-Hāmilī

Died in 765/1363-64 (*HK*, II, p. 24) or 769/1367-68 (*HK*, V, p. 454).

References: *Suter*, no. 260; *Brockelmann* II, p. 236 and SII, p. 240.

4.1 Maʿūnat al-ṭullāb fī maʿrifat al-ḥisāb

A commentary on al-Ṣardafī's treatise on arithmetic (see 1.1 above) compiled in 724/1324, mentioned by Ḥājjī Khalīfa (*HK*, II, p. 24 and V, p. 454): MS Berlin Ahlwardt 5977 (cop. ca. 900/1494).

5. Aḥmad ibn Mūsā ibn ʿAlī al-Jallād al-Nakhlī (?) al-Faraḍī

Born 700/1300-01, died 792/1390, father of no. 6 below. Wrote on inheritance, arithmetic and geometry, for the Sultan al-Ashraf Ismāʿīl (see 18a in Part II).[1] In the colophon of the Hyderabad manuscript listed below it is stated that he lived from 755 Hijra (perhaps this should be 700 Hijra) to 792 Hijra and that he studied under his father, who in turn had

[1] *Löfgren & Traini*, p. 143.

studied inheritance and arithmetic under ʿAlī ibn ʿAbd Allāh al-Faraḍī al-Zīlʿī, who was a student of the Imam Aḥmad ibn Mūsā al-ʿUjayl.

5.1 al-Muqaddima al-durrīya fī stinbāṭ al-ṣināʿa al-jabrīya

A short treatise on algebra: MS Hyderabad Osmania University Library 1549/520-J-MD (4 fols., 1061 Hijra).

6. Alī ibn Aḥmad ibn Mūsā al-Jallād al-Faraḍī

A son of no. 5 above, born 732/1331-32.[2]

6.1 Lubb al-lubāb fī ṭarāʾiq al-ḥisāb

On the arithmetic of inheritance in two *faṣl*s: MSS Milan Ambrosiana CCLXXI,A (fols. 1r-31v, 794 Hijra), unique.

7. Abū ʿAbd Allāh Muḥammad ibn ʿAbd Allāh ibn ʿAbd al-Raḥmān ibn Salm

References: *Brockelmann*, SII, p. 855.

7.1 A commentary on the arithmetic of al-Ṣardafī (see no. 1 above): MS Ambrosiana D550.

8. ʿAbd Allāh ibn ʿUmar ibn Khalīl al-Yamanī

References: *Azzawi*, p. 338.

Azzawi states that he was a pious scholar of mathematics, but gives no further information.

9. Al-Jaḥḥāf

References: *Brockelmann*, SII, p. 567; *Sayyid*, pp. 256-257; see also no. 39 in Part II.

9.1 Al-Ṭarīqa al-jalīla = Ṭariqat al-ḥussāb fī ṣināʿat al-kuttāb = Ṭarīqat Jaḥḥāf

MSS Cairo K4309 (fols. 1r-85v, cop. 1222 Hijra), entitled *al-Ṭarīqa al- aẓīma*); Vatican V.1047,4; Vatican V.1078,7 (where the name is Sharaf al-Islām Ḥasan ibn Shams al-Dīn

[2] *Lofgren & Traini*. p. 143.

al-Jaḥḥaf); Cairo Dār al-Kutub *majāmī*ᶜ 705,11 (fols. 67v-84v) (where the name is Shams al-Dīn Ibrāhīm ibn Mahdī).

9.2 Sharh Miftāḥ al-Fā'iḍ fī ᶜilm al-farā'iḍ

MS Vatican I.404,11.

10. Aḥmad al-Ḥusaynī al-Yamanī (?)[3]

10.1 Risāla fī Bayān ḍābiṭat ᶜuqūd al-aᶜdād

A treatise on arithmetic in five *faṣl*s: MSS Istanbul Esad Ef. 3673,5 (fols. 27-28); Cairo Ṭalᶜat *majāmī*ᶜ 635,10 (fols. 54v-56r, 1170 Hijra).

11. ᶜAfīf al-Dīn ᶜAbd Allāh ibn Muḥammad ibn Ibrāhīm ibn ᶜAṭīya ibn Muḥammad ibn Aḥmad ibn Muḥiy al-Dīn al-Ḥārithī al-Najrānī al-Madanī al-Madhḥījī

References: *Brockelmann*, SII, p. 253.

11.1 Al-Riyāḍ al-naffāḥa fī ᶜilm al-misāḥa

On surveying: MS Ambrosiana B16,1 (cf. *RSO*, IV, p. 96).

12. Anonymous

12.1 Al-Risāla al-Yamanīya fi l-ḥisāb

A treatise on arithmetic arranged in twelve *faṣl*s: MS Cairo Dār al-Kutub *majāmī*ᶜ 713,10 (fols. 104v-113r, ca. 1000 Hijra). This manuscript is of Yemeni provenance.

[3] The appellation al-Yamanī occurs only in the Istanbul manuscript.

APPENDIX B

SOME NEW MATERIAL

After this study was completed, and whilst I was attending the Second International Symposium for the History of Arabic Science in Aleppo in 1979, I met Mr. ʿAbd Allāh Muḥammad al-Ḥibshī of Sanaa, author of a new book *Maṣādir al-fikr al-ʿArabī al-Islāmī fi l-Yaman*, published for the Centre of Yemeni Studies in Sanaa by Dār al-ʿAwda, Beirut, 1979. Mr. Ḥibshī showed me the sections on astronomy and mathematics in his book, and I am glad to be able to incorporate some additional information gathered by Mr. Ḥibshī. With the possible exception of references to a Yemeni adaptation of the *Zīj* of the fourteenth-century Syrian astronomer Ibn al-Shāṭir and a treatise by al-Ṣarmī, most of this material relates to folk astronomy and calendrical tables.

p. 483: AL-NUʿMĀNĪ, 5th century H, author of *Qaṣīda bāʾīya fī dhikr al-shuhūr wa-l-kurūm* . . . , a poem on the months of the (solar) year: manuscript in the Library of Mushrif b. ʿAbd al-Karīm (location?), photographed by the Arab League Institute of Arabic Manuscripts.

p. 483: *ad* AL-AṢBAḤĪ (see my Section 5.1) add MS Sanaa Grand Mosque Library *falak* 34 (copied 746H) of his treatise *al-Yawāqīt* (??).

p. 483: *ad* AL-FĀRISĪ (see my Section 6), add MS Medina *majāmīʿ* 25 of the *Nihāyat al-idrāk* and the *Maʿārij al-fikr*.

p. 484: Ḥibshī lists ISMĀʿĪL B. AḤMAD AL-NAJRĀNĪ, with his date of death as 794 Hijra (cf. p. 376), but this is not the astronomer al-Najrānī. He also lists a *Risāla fi ʿIlm al-nujūm wa-l-zījāt*, without further information.

p. 484: AL-ḤASAN B. ḤUMAYD AL-MIQRĀʾĪ, d. after 850 Hijra, author of *Risāla al-Fawāʾid wa-l-asrār fīmā yataʿallaq bi-l-falak al-dawwār*, no reference given (though see p. 120).

p. 484: ʿABD ALLĀH B. ʿABD AL-RAḤMĀN BA-FAḌL, d. 918 Hijra, author of *Risāla fī ʿIlm al-falak* and *Risāla fī Maʿrifat samt al-qibla*, no reference given (though see p. 284).

p. 484: MUḤAMMAD B. ʿUMAR BAḤRAQ (?), d. 930 Hijra, author of *Risāla fī ʿIlm al-mīqāt*, no reference given (though see p. 123).

pp. 484-485: *ad* BĀ MAKHRAMA, add biographical information including b. 907 Hijra, d. Aden, 972 Hijra, and titles as follows: *al-Lumʿa fī ʿilm al-falak* in MS Rabat 3023 Kattani, and *al-Jadāwil al-muḥaqqaqa fī ʿilm al-hayʾa* (no reference); *Risāla fī Samt al-qibla* (no reference); *Risāla fī l-Rubʿ al-mujayyab* in a MS in the Library of Āl al-Bār in al-Qarn Dūʿan (?); *Risāla fī Maʿrifat al-awqāt wa-l-sāʿāt* (no reference); *Risāla fi Ikhtilāf al-maṭāliʿ wa-ttifāqihā* (no reference–cf. Section 26.2 in Part II).

p. 485: MUḤAMMAD B. AL-ṢIDDĪQ AL-KHĀṢṢ, d. 996 Hijra (see Section 41.1 in Part II), author of *Jadāwil fī ʿilm al-falak* in the Ambrosiana (no number given). No further details (though see p. 25).

p. 485: ABU SAʿĪD AḤMAD B. MUḤAMMAD AL-SHAJARĪ, author of *al-Dalāʾil fī aḥkam al-nujūm* in MS Ambrosiana G 170.

p. 485: ʿABD AL-QAYYŪM AL-RAGHĪLĪ, d. 1046 Hijra, author of *al-Mudkhal al-mukhtaṣar li-Zīj Ibn al-Shāṭir al-musammā bi-l-Durr al-naẓīm* (no manuscripts). (The *Zīj* entitled *al-Durr al-naẓīm* was itself a recension of the *Zīj* of Ibn al-Shāṭir by a later author.)

p. 485: MUḤAMMAD B. AḤMAD B. ʿIZZ AL-DĪN known as IBN AL-ʿANZ (?), d. 1053 Hijra, author of a commentary (no manuscripts mentioned) on a poem on timekeeping by al-Hādī ʿIzz al-Dīn (see my Section 25.1), no manuscripts mentioned.

pp. 485-486: Ḥibshī mentions MUḤAMMAD B. ABĪ BAKR B. AḤMAD AL-SHILLĪ, b. Hadramaut, 1030 Hijra, d. Mecca, 1093 Hijra, author of *Risāla fi l-Rubʿ al-mujayyab*; *Risāla fī ʿIlm al-mīqāt*; *Risāla fī Maʿrifat ẓill al-zawāl*; *Risāla fī Maʿrifat ittifāq al-maṭāliʿ wa-khtilāfihā*; *R. fi l-Muqanṭar*; and *R. fi l-Asṭurlāb*. No manuscripts are mentioned. However, the third title probably relates to a short treatise on the midday shadows at latitude 21°, Mecca, preserved in MS Cairo Dār al-Kutub *mīqāt* 130,2 (fols. 14v-15v, ca. 1200 Hijra), and summarized in *SATMI*, III. Likewise, al-Shillī's writings on quadrants would relate him to the Egyptian-Hejazi tradition rather than the Yemeni tradition. On al-Shillī see further *Brockelmann*, II, p. 502, and SII, p. 516.

p. 486: ʿABD ALLĀH B. ṢALĀḤ ʿANQŪB, 11th century Hijra, author of *Majmūʿ al-zīj*; and *Takmīl al-shuhūr al-Yazdigirdīya* up to 1168 Hijra. No references cited.

p. 486: BĀGHŪTH AL-ḤAḌRAMĪ, author of a *Kitāb fi l-Falak*. No manuscripts mentioned.

p. 486: Ḥibshī lists MAḤẒŪẒ B. ʿABD AL-RAḤMĀN BĀ ʿAYYĀD, author of *al-Shāmil* . . . , in MS Sanaa Grand Mosque *majāmīʿ* 74. See Section 26.1 in Part II.

pp. 486-487: ISḤĀQ B. MUḤAMMAD AL-ʿABDĪ, d. 1115 Hijra (on whom see also p. 133), author of *Risālat Awāʾil al-shuhūr al-ʿArabīya wa-mawāqīt al-ṣalāt* extant in MS Cairo Dār al-Kutub *falsafa* 411 (copied 1180 Hijra).

p. 487: Ḥibshī lists HĀDĪ B. ʿALĪ AL-ṢARMĪ, d. 1121 Hijra, author of *Shams al-awān fī-mā taʿāqaba ʿalayhi al-mulwān* (?), apparently an important work on astronomy (no manuscripts recorded).

p. 487: Ḥibshī lists ḤASAN B. ZAYD B. ʿALĪ B. IBRĀHĪM JAḤḤAF (see Section 39 in Part II and Appendix A, no. 9), giving biographical information including the date of his birth 1094 Hijra, and of his death in 1127 Hijra in Sanaa, and recording the existence of a *Risāla fī ʿIlm al-mīqāt wa madākhil al-shuhūr* . . . in MS Sanaa Grand Mosque *majāmīʿ* 64. This information adds yet more confusion to our knowledge of Jaḥḥāf!

p. 487: ʿALĪ B. ḤASAN AL-AKWAʿ , d. 1203 Hijra, author of a calendrical table preserved in MS Ambrosiana G5.

p. 487: ʿUMAR B. SAQQĀF B. MUḤAMMAD AL-SAQQĀF, d. 1216 Hijra, author of *al-Maṭālib al-sunnīya/al-sanīya (?) fi l-fawāʾid al-falakīya*, no references cited (though see p. 299).

pp. 487-488: MUḤAMMAD B. AḤMAD B. AL-MANṢŪR AL-ḤUSAYNĪ AL-ḤUSNĪ, b. Sanaa, 1163 Hijra, d. 1217 Hijra, author of some calendrical tables (no manuscripts listed).

p. 488: MUḤAMMAD B. ʿABD AL-RAḤMĀN B. SULAYMĀN B. YAḤYĀ AL-AHDAL, d. ca. 1260 Hijra (see also p. 391), author of *Rafʿ al-ishtibāh fī masāʾil al-quṭb wa-l-jāh*, on the celestial pole and the pole star, extant in MS Riyadh University 1517 (copied 1239 Hijra).

p. 488: YAḤYĀ B. MAẒHAR (or MUṬAHHAR?) B. ISMĀʿĪL, d. 1268 Hijra (see also pp. 72-73), author of *Jadwal fī ʿilm al-falak*, no manuscripts mentioned.

p. 488: *ad* ABD ALLĀH B. ḤAMZA AL-DAWWĀRĪ (see Section 41.1, no. 14 in Part II), add MS Sanaa Grand Mosque Western Library *majāmīʿ* 98 of his *Bulghat al-muqtāt*; and two other astrological works, *Malḥama tunbaʾ fīhā bimā sayakūn* and *Maʿdan al-jawāhir fī ikhrāj al-ḍamāʾir*, no manuscripts mentioned.

p. 488: ʿABDʿALLAH B. AHMAD AL-KHAYRĪ AL-SHAMĀKHĪ, *fl.* 13th century Hijra, author of a calendrical table entitled *al-Jadwal al-thamīn*, extant in MS Ambrosiana G20 (?).

p. 488: MUḤAMMAD B. ḤĀMID B. ʿUMAR AL-SAQQĀF, d. 1338 Hijra (see pp. 251-252), author of *Naṣb al-shark fi qtināṣ mā yuḥtāj ilayhi min ʿilm al-falak*, no manuscripts recorded.

p. 489: LUTF ALLAH B. ABD ALLAH B. ʿABD ALLAH B. HAMZA AL-DAWWARI, d. 1349 Hijra, continued the calendrical tables of his grandfather (see above), no manuscripts listed.

p. 494: ʿABD AL-QĀDIR B. AḤMAD AL-KAWKABĀNĪ, d. 1207 Hijra, author of *Risāla fi l- ʿAmal fi-l-ḥisāb al-qaṭʿī idhā khālafa stināduhu ʿalā ruʾyat al-hilāl*, no manuscripts listed.

The information on Yemeni authors on arithmetic, surveying, and inheritance given in Appendix A above can be supplemented with the information given by Ḥibshī on pp. 490-494 and pp. 259-268.

SIGLA OF MANUSCRIPTS CONSULTED

nota bene: The sigla are the same as those used in *SATMI*, I and II with numerous additions. An asterisk denotes that I have not inspected a given manuscript or a copy thereof, but have relied on the description in the appropriate catalogue.

For manuscript catalogues see *Huisman* and *Sezgin*, VI. Griffini's unfinished catalogue of the Ambrosiana collection of Yemeni manuscripts has recently been completed by Prof. O. Löfgren of Uppsala. Whilst this study was being submitted for publication, the first volume of this new catalogue was published (see *Löfgren & Traini* in the bibliography). Cairo manuscripts are listed in my catalogue of the scientific manuscripts in the Egyptian National Library (see *Cairo Cat.*).

The section of the present study in which the manuscripts is discussed is indicated in parentheses. Yemeni mathematical manuscripts are listed in the appendix above.

AK*	MS Algiers Fagnan 1485,3 al-Thābitī's prayer-tables for the Yemen	II.33.1
AL	MS Alexandria Municipal Library 5577C *Taysīr al-maṭālib fī tasyīr al-kawākib* (Zīj) by Muḥammad ibn Abī Bakr al-Kawāshī	I.2; II.7.1; II.38.0
AR	MS Alexandria Municipal Library 3010D *Maʿārij al-fikr al-wahīj* by Muḥammad ibn Abī Bakr al-Fārisī	II.6.2
AZ*	MS Baghdad Awqāf 2982/6276 *al-Yawāqīt fi l-mawāqīt* by Ibrāhīm ibn ʿAlī ibn Muḥammad al-Aṣbaḥī al-Yamanī	I.3, note 13; II.5.1
BG*	MS Berlin Ahlwardt 5711/5883/5893/5898 (= Landberg 221) Yemeni copy of (1) the astrological treatise of Abu l-ʿAnbas al Ṣaymarī (2) an astrological treatise by Sahl ibn Bishr al-Isrāʾīlī (3) A fragment from the astrological treatise *al-Kāfī* by Abū ʿAlī al-Marrākushī (4) An anonymous astronomical treatise and the *Kitāb fi Qiyām al-khulafāʾ* of Māshāʾallāh	I.2, note 16
BM*	MS Berlin Ahlwardt 5731 (= Glaser 163) [*Tuḥfat al-rāghib* by al-Fārisī] (incomplete)	II.6.1

BN	MS Berlin Ahlwardt 5784/5769/5720 (= Mq. 733)	I.2; II.38.0
	(1) *al-Yawāqīt fī maʿrifat al-mawāqīt* (almanac) by al-Ḥusayn ibn Zayd ibn ʿAlī Jaḥḥāf (fols. 1-7r)	II.39.1
	(2) al-Thābitī's prayer-tables for the Yemen (fols. 8v-15v)	II.33.1
	(3) Anonymous corpus of tables for timekeeping computed for Taiz [Abu l-ʿUqūl's *Mirʾāt al-zamān*] (fols. 16r-166r)	II.9.3; II.38.2
	(4) Tables for timekeeping computed for Sanaa by ʿAbd Allāh ibn Ṣalāḥ Dāʿir (fols. 121v, 124r-127r)	II.31.1
BR	MS Berlin Ahlwardt 5664	I.2; II.2.1
	Treatise on folk astronomy by Muḥammad ibn Raḥīq ibn ʿAbd al-Karīm	
BW*	MS Berlin Ahlwardt 5746/5768 (= Glaser 21)	
	(1) A poem on the lunar mansions by al-Mahdī Aḥmad ibn Yaḥyā	II.24a.1
	(2) Calendrical tables by Muḥammad ibn Aḥmad al-Najjār al-Anṣarī	II.41.1(13)
	(3) Calendrical tables by Muḥammad ibn Aḥmad ibn al-Imām	II.41.1(12)
BX*	MS Berlin Ahlwardt 5669 (= Landberg 33)	I.2, note 16
	Yemeni copy of al-Kharaqī's *Muntaha l-idrāk* (copied ca. 650H)	
BY*	MS Berlin Ahlwardt 5789 (= Glaser 176)	II.41.1(17)
	Anonymous calendrical tables	
BZ*	MS Berlin Ahlwardt 5826 (= Glaser 227)	I.2, note 16
	Yemeni copy of the treatise on the use of the sine quadrant by Yaḥyā ibn Muḥammad al-Ḥaṭṭāb	
CZ	MS Cambridge University Library Gg. 3.27	I.2
	(1) Anonymous recension of al-Fārisī's *al-Zīj al-Mumtaḥan al-Khazāʾinī*	II.21.1
	(2) *al-Zīj al-Mumtaḥan al-Khazāʾinī* by Muḥammad ibn Abī Bakr al-Fārisī	II.6.3
DA	MS Damascus Ẓāhirīya 3092	I.2
	(1) *al-Zīj al-Majmūʿ*, anonymous (fols. 1r-73r)	II.35.1; II.38
	(2) Planetary tables by Ismāʿīl ibn ʿAṭīya al-Najrānī (fols. 77v-84r)	II.23.1
	(3) Anonymous planetary equation tables (fols. 85v-98r)	

DP	MS Dublin Chester Beatty 4562 *Nihāyat al-idrāk* by Muḥammad ibn Abī Bakr al-Fārisī	II.6.4
DT	MS Dublin Chester Beatty 4090 *Kitāb Tatmīm ʿamal al-asṭurlāb* by al-Bakhāniqī	II.13.1
HD	Hyderabad Asafiya 395 *Ghāyat itqān al-ḥarakāt* by al-Sarḥī	II.37.1
HE	Hyderabad Salar Jung Museum astronomy 29	I.2, note 16
	(1) Introduction to astrology by Abu l-Qāsim al-Balkhī	
	(2) *Ghāyat itqān al-ḥarakāt* by al-Sarḥī	II.37.1
	(3) Simple sexagesimal multiplication table	
	(4) Tables of the astrological houses for the latitude of Sanaa	II.29.1
JA	MS in the private collection of Sayyid Luṭf Yūsuf al-Mutawakkil, Jibla, Yemen	
	(1) *Maʿārij al-fikr al-wahīj* by Muḥammad ibn Abī Bakr al-Fārisī	II.6.2
	(2) *Kitāb al-Sirr al-maktūm* by Fakr al-Dīn al-Rāzī	
LA	MS London British Library Or. 3624 (Rieu 768) *al-Zīj al-Mukhtār* by Abu l-ʿUqūl	I.2; II.9.1
LE	MS London British Library Or. 3748 (Rieu 769) *Ghāyat itqān al-ḥarakāt (Zīj)* by ʿAbd Allāh ibn ʿAbd Allāh al-Sarḥī	II.37.1
LF	MS London British Library Or. 3849 (Rieu 770)	
	(1) Anonymous calendrical tables (from 1201 to 1300 Hijra) (= MS LG)	II.41.1(10)
	(2) Simple shadow tables	II.40.1
LG	MS London British Library Or. 3713 (Rieu 771) Calendrical tables (from 1201 to 1300 Hijra) by Jamāl al-Dīn ʿAlī ibn al-Ḥasan ibn Muḥammad al-Akwaʿ	II.41.1(10)
LH	MS London British Library Or. 3732 (Rieu 772)	
	(1) Calendrical tables (from 1215 to 1230 Hijra) by Muḥammad ibn Aḥmad ibn al-Imām	II.41.1(12)
	(2) Simple shadow tables	II.40.1
LI	MS London British Library Or. 3747 (Rieu 773)	
	(1) Calendrical tables (from 1263 to 1300 Hijra) by by Fakhr al-Islām ʿAbd Allāh ibn Ḥamza	II.41.1(14)

	(2) *al-Yawāqīt fī maʿrifat al-mawāqīt* by Abu l-ʿUqūl	II.9.4
	(3) A poem on the solar months and the foods that should be eaten during them, by al-Yāfiʿī	II.15.2
	(4) Anonymous treatise on the lunar mansions entitled *Kitāb al-Īḍāḥ al-shāfī bi-l-itqān fī maʿrifat al-manāzil wa-l-zamān*	II.32.1
LJ	MS London British Library Or. 3848 (Rieu 774)	
	(1) Anonymous calendrical tables (from 1181 to 1253 Hijra)	II.41.1(8)
	(2) Treatise on the seasons	
	(3) Calendrical tables (from 1257 to 1301 Hijra) by ʿAbd Allāh ibn Ḥamza	II.41.4(14)
	(4) Prayer-tables	II.40.1
	(5) Anonymous calendrical and Ottoman-type prayer tables for 1293 Hijra	II.40.1
LK	MS London British Library Or. 3906 (Rieu 1226) *al-Hayʾa al-sanīya fi l-hayʾa al-sunnīya* by Jalāl al-Dīn al-Suyūṭī (eighteenth century Yemeni copy)	I.2, note 16
LM	MS London British Library Or. 9116 *Taysīr al-maṭālib fī tasyīr al-kawākib* (*zīj*) (anonymous)	II.7.1
LQ*	Leiden Or. 2807(1) (104 pp.?) *Mufīdat al-sāʾil* by Yaḥyā ibn Muḥsin ibn Aḥmad ibn Rājiḥ	Addenda (4)
MA	MS Milan Ambrosiana C46, fols. 52-58 *al-Yawāqīt fī maʿrifat al-mawāqīt* by Abu l-ʿUqūl	II.9.4
MB	MS Milan Ambrosiana C84 Anonymous tables for timekeeping for Taiz entitled *Tuḥfat al-awān al-muntazaʿa min Mirʾāt al-zamān* [extracted from the *Miraʾāt al-zamān* by Abu l-ʿUqūl]	I.2; II.6.1; II.9.3
ME	MS Milan Ambrosiana X 73 sup. (Griffini 37) *Tuḥfat al-rāghib* by Muḥammad ibn Abī Bakr al-Fārisī	I.2; II.6.1
MF*	MS Milan Ambrosiana CCLIII (= Y 204 sup.) Calendrical tables by Saʿīd ibn ʿUmar Bā Bashīr al-Shdy (?)	II.41.1(1)
MG*	MS Milan Ambrosiana CCXCIV (= X 235 sup = Griffini 47), F (fols. 114-123) Calendrical tables by Muhammad al-ʿAṭṭās ibn Sallām Bā Faḍl al-Ḥaḍramī al-Shāfiʿī	II.41.1(3)

MH	Milan Ambrosiana C82 Maghribi recension of the perpetual almanac of Zacuto (Yemeni copy)	I.2, note 16; II.38.0
MR	MSS Milan Ambrosiana C79, C161, D21, D252, D268, D365, E100, E170, F67, F145, F267 Eleven Yemeni manuscripts containing simple calendrical tables	II.41.1 (5 & 8 & 14)
MT	MSS Milan Ambrosiana E16, E403, F201, F202 Four copies of the *Ghāyat itqān al-ḥarakāt* (*zīj*) by ʿAbd Allāh ibn ʿAbd Allāh al-Sarḥī	II.37.1
MU	MS Milan Ambrosiana C83 Almanac for the solar year 1733-34 by Yūsuf ibn Yūsuf al-Maḥallī	II.38.1
MV	MS Milan Ambrosiana C86 (1,2) Astrological treatises attributed to Hermes, and notes	
	(3) al-Fārisī's translation of *al-Aḥkām al-Jāmaspīya* (17 folios)	II.6.7
	(4) Ahmad ibn Yūsuf al-Miṣrī's commentary on Ptolemy's *Centiloquium*	
	(5) Astrological notes by ʿIzz al-Dīn ibn Aybak al-Amhadī (?)	
	(6) Fragments from the *Zīj*es of al-Battānī and al-Najrānī in considerable confusion (4 folios)	I.2, note 16; II.23.1; II.24.0
NE	MS Istanbul Nurosmaniye 2951,1 (fols. 1v-32r) *Maʿārij al-fikr al-wahīj* by Muḥammad ibn Abī Bakr al-Fārisī	II.6.2
NP	MS Istanbul (Millet Genel) Ali Emiri 2722 *Nihāyat al-idrāk* by Muḥammad ibn Abī Bakr al-Fārisī	II.6.4
NQ	MS Istanbul (Suleymaniye) Hosrev Paşa 216 *Nihāyat al-idrāk* by Muḥammad ibn Abī Bakr al-Fārisī	II.6.4
ON	MS Oxford Bodleian Huntington 233 (Uri 905) *al-Tabṣira fī ʿilm al-nujūm* by the Sultan al-Ashraf ʿUmar ibn Yūsuf	I.2; II.8.1
PA	MS Paris Bibliothèque Nationale ar. 2523 Anonymous Yemeni *Zīj* compiled in Taiz	I.2; II.17.1

RA	MS Cairo DM (= Dār al-Kutub *mīqāt*) 145 (Arab League Institute of Arabic Manuscripts, Microfilm no. *falak* 189) *Ma'ārij al-fikr al-wahīj* by Muḥammad ibn Abī Bakr al-Fārisī	II.6.2
SA	MS in the private collection of Qadi Ismail al-Akwa, Sanaa *al-Zīj al-Mumtaḥan al-Khazā'inī* by Muḥammad ibn Abī Bakr al-Fārisī (lacks introduction)	I.2; II.6.3
SB	MS in the private collection of Qadi Ismail al-Akwa, Sanaa Compendium containing astronomical tables, apparently compiled by the Sultan al-Afḍal al-'Abbās ibn 'Alī	I.2; II.9; II.10.1; II.12.1; II.18.1
SC	MS in the private collection of Qadi Ali Muhammad al-Sharafī, Sanaa *Ghāyat itqān al-ḥarakāt* (*Zīj*) by 'Abd Allāh ibn 'Abd Allāh al-Sarḥī	II.37.1
SD	MS in the private collection of Qadi Muhammad Ali al-Sharafī, Sanaa *Ghāyat itqān al-ḥarakāt* (*Zīj*) by 'Abd Allāh ibn 'Abd Allāh al-Sarḥī	II.37.1
SE	MS in the private collection of Shaykh Ahmad Muhammad al-Hatimi, Sanaa *Bughyat al-ṭālib* (*Zīj*) by al-Ḥasan ibn 'Abd Allāh al-Sarḥī	II.36.1
SF	MS Sanaa Grand Mosque Library, *falak* 492 *al-Zīj al-Mumtaḥan al-Khazā'inī* by Muḥammad ibn Abī Bakr al-Fārisī	II.6.3
SG	MS Sanaa Grand Mosque Library, unnumbered	I.2
	(1) *Ghāyat itqān al-ḥarakāt* (*Zīj*) by 'Abd Allāh ibn 'Abd Allāh al-Sarḥī (fols. 1r-73r)	II.37.1
	(2) Fragment from an anonymous Sanaa *Zīj* (fols. 75v-81r)	II.16.1
SH	MS Sanaa Grand Mosque Library, *falak* 14 *Ghāyat itqān al-ḥarakāt* (*Zīj*) by 'Abd Allāh ibn 'Abd Allāh al-Sarḥī	II.37.1
SI	MS Sanaa Grand Mosque Library, unnumbered Anonymous Ottoman-type prayer-tables for Sanaa	II.40.1
SJ	MS Sanaa Grand Mosque Library, *falak* 2 (?)	
	(1) *Zahr al-zuhūr* (calendrical tables) by 'Abd al-Wāsi' ibn Yaḥyā al-Wāsi'ī	II.41.1(16);

	(2) *Tuḥfat al-thiqāt* (calendrical tables) by ʿAbd al-Wahhāb ibn ʿAlī	II.41.1(15)
SK	MS Sanaa Grand Mosque Library, *mājamīʿ* 27, fols. 3r-19v	
	(1) al-Thābitī's prayer-tables for the Yemen (fols. 3r-7v)	II.33.1
	(2) *Bulghat al-muqtāt* (calendrical tables) by ʿAbd Allāh ibn Ḥamza (fols. 8r-19v)	II.41.1(14)
SL	MS in the private collection of Qadi Husayn al-Sayyaghi, Sanaa (Photographs are preserved as MS Cairo Dār al-Kutub K7012.) al-Hamdānī's *Sarāʾir al-ḥikma* (ch. 10 only)	I.2; II.1.4
SM	MS Sanaa Grand Mosque Library, *majāmī* 23 (Photographs preserved as MSS Cairo Dār al-Kutub K4678 and K16000.) *ʿĀdāt al-nujūm* by Abu l-Ghanāʾim Muslim ibn Maḥmūd al-Shayzarī	II.4.1
SN	MS Sanaa Grand Mosque Library 101 *al-Yawāqīt fī tahqīq al-mawāqīt* by Muḥammad ibn Ismāʿīl	
SP	MS Sanaa Grand Mosque Western Library, unnumbered (Arab League N. Yemeni microfilms no. 285) Treatise on the calendar attributed to Jaʿfar al-Ṣādiq and an anonymous almanac for the Syrian months, probably of Yemeni origin.	Addenda (2)
SQ	MS Princeton Mach 5015 = Yehuda 4224, fols. 113v-117r al-Yāfiʿī's poem on the Syrian months	II.15.2
SR	MS Leiden Landberg-Brill $^{1}78$, $^{2}141_{2}$ (*Brockelmann*) (now in Princeton?) *Sirāj al-tawḥīd* . . . by al-Yāfiʿī	II.15.1
TA	MS Cairo TR (= Taymūr *riyāḍiyāt*) 105 *Muʿīn al-ṭālib fī l-ʿamal bi-l-asṭurlāb* by the Sultan al-Ashraf ʿUmar ibn Yūsuf	I.2; II.8.2
TC	MS Cairo TR (= Taymūr *riyāḍiyāt*) 274 (Photographs are preserved in MS Cairo Dār al-Kutub K7198) Anonymous almanac for Taiz, 808 Hijra (= 1405/6)	I.2; II.22.1
TF	MS Cairo TR (= Taymūr *riyāḍiyāt*) 227,1 *Maʿārij al-fikr al-wahīj* by Muḥammad ibn Abī Bakr al-Fārisī	II.6.1

TG MS Cairo DM (= Dār al-Kutub *mīqāt*) 817 I.2

(1) *Maᶜārij al-fikr al-wahīj* by Muḥammad ibn Abī Bakr al-Fārisī II.6.1

(2) Anonymous almanac for Taiz, 727 Hijra (= 1326-27) II.11.1; II.22.1

TH MS Cairo DM (= Dār al-Kutub *mīqāt*) 948

(1) *al-Yawāqīt fi l-mawāqīt* by Ibrāhīm ibn ᶜAlī ibn Muḥammad al-Aṣbaḥī I.2; II.5.1

(2) *Kitāb al-Shāmil fī dalāᵓil al-qibla wa-ḥisāb al-Rūm wa-l-manāzil* attributed to Bā Makhrama II.26.1

(3) *Nubdha fī . . . mā yataᶜallaq bi-waqt al-ṣalāt wa-l-qibla* attributed to Abu l-Qāsim II.42.1

(4) Notes on the lunar mansions

(5) Notes on shadow lengths corresponding to the Syrian calendar

(6) Miscellaneous notes

(7) Notes on the shadow lengths at Aden and Taiz by ᶜAfīf al-Dīn ᶜAbd Allāh ibn ᶜUmar Bā Makhrama II.26.3

(8) Poem on shadow lengths by ᶜAfīf al-Dīn ᶜAbd al-Wālī al-ᶜIrāqī

TI MSS Cairo DM (= Dār al-Kutub *mīqāt*) 180, 191, 192, 983, 1196 and ṬM (= Ṭalᶜat *mīqāt*) 157,2, 248,1 II.6.4

Seven copies of *Nihāyat al-idrāk* by Muḥammad ibn Abī Bakr al-Fārisī

TJ MS Berlin Ahlwardt 5888 (= Sprenger 1873) II.6.4

Nihāyat al-idrāk by Muḥammad ibn Abī Bakr al-Fārisī

TK MS Cairo MḤ (= Dār al-Kutub Muṣṭafā Fāḍil *ḥurūf*) 73,2

Ayyāt al-āfāq fī khawāṣṣ al-awfāq by Muḥammad ibn Abī Bakr al-Fārisī

TL MS Cairo DM (= Dār al-Kutub *mīqāt*) 1015 II.19.1

Madkhal al-taᶜlīm fī inshāᵓ al-tasyīr wa-ᶜamal al-taqwīm by Abū Bakr ibn Abi l-Maᶜālī

TZ* MS Tehran Majlis al-Umma al-Īrānī vol. 2, p. 82 (referred to in *Azzawi*, p. 234) II.8.2

Manhaj (?) al-ṭullāb fi l-ᶜamal bi-l-asṭurlāb by the Sultan Ashraf ᶜUmar ibn Yūsuf

VB	MS Vatican ar. 962	
	(1) Calendrical tables (from 1177 to 1200 Hijra) computed by (?) Jamāl al-Dīn ʿAlī ibn Muḥammad ibn Ḥusayn al-Murtaḍā (fols. 5r-12v)	II.41.1(7)
	(2) al-Thābitī's prayer-tables for the Yemen (fols. 13r-19r)	II.33.1
	(3) Anonymous calendrical tables (from 1183 to 1253 Hijra) (fols. 20v-58r)	II.41.1(9)
	(4) Calendrical tables (from 1201 to 1325 Hijra) by ʿAfīf al-Dīn ʿAbd Allāh ibn Aḥmad al-Khayrī al-Shatmakhī (?) al-Zabīdī (fols. 63v-70v)	II.41.1(11)
	(5) Calendrical tables (from 1197 to 1300 Hijra) by Ḥusayn ibn Zayd Jaḥḥāf (fols. 76v-84r)	II.39.2; II.41.1(6)
	(6) Extract from al-Thābitī's prayer-tables for the Yemen (fols. 84v-88v)	
VD	MS Vatican ar. 1038, fols. 4v-25r Anonymous ephemerides for Sanaa, 1146 Hijra	II.38.1
VE*	MS Vatican ar. 955, fols. 1-89v *Ghāyat itqān al-ḥarakāt* (*Zīj*) by ʿAbd Allāh ibn ʿAbd Allāh al-Sarḥī	II.37.1
VF	MS Vatican ar. 964	
	(1) Anonymous calendrical tables (from 1201 to 1300 Hijra) (fols. 1r-11v)	II.41.1(10)
	(2) Extract from al-Thābitī's prayer-tables (fols. 12r-14v)	
	(3) Anonymous calendrical tables (from 1101 to 1200 Hijra) based on those of al-Ṣiddīq ibn Muḥammad al-Khaṣṣ (fols. 16v-27v)	II.41.1(2)
	(4) Anonymous calendrical and astrological fragments (fols. 28r-31r)	
	(5) Anonymous ephemerides for Cairo, 1067 Hijra (fols. 31v-52v)	
VG*	MS Vatican ar. 1120, fols. 61-63 *al-Yawāqīt fī maʿrifat al-mawāqīt* by Abu l-ʿUqūl	II.9.4
VK*	MS Vatican ar. 949 Contains a fragment of a commentary on the *Urjūza fi l-shuhūr al-Rūmīya* by Nashwān ibn Saʿīd al-Ḥimyarī (*Serjeant* 2, p. 174)	II.3.2
YA	MS Cairo DM (= Dār al-Kutub *mīqāt*) 400 *al-Zīj al-Balīgh* by Kūshyār ibn Labbān (with annotations by a Yemeni astronomer)	I.2, note 16

YB*	MS Gotha Forschungsbibliothek A1523 Fragment of *Kitāb Tatmīm ʿamal al-asṭurlāb* by al-Bakhāniqī (one folio)	II.13.1
YC*	MS Istanbul (Suleymaniye) Hamidiye 830,2 *Nihāyat al-idrāk* by Muhammad ibn Abī Bakr al-Fārisī	II.6.4
YD	MS Istanbul Topkapi 7098 (H 466) Late Egyptian copy of al-Fārisī's *Nihāyat al-idrāk*	II.6.4
YE	MS Paris Bibliothèque Nationale ar. 6840 al-Bīrūnī's *al-Qānūn al-Masʿūdī* (copy owned by the Yemeni Imām ʿAbd Allāh, *fl.* ca. 1820)	I.2, note 16
YF	MS Oxford Bodleian Marsh 134, fols. 9r-52r *al-Yawāqīt fi l-mawāqīt* by Ibrāhīm ibn ʿAlī ibn Muḥammad al-Aṣbaḥī	II.5.1
YG	MS Cambridge University Library Arberry Supp. 110(a) = Or. 1236(11) *al-Yawāqīt fi l-mawāqīt* by Ibrāhīm ibn ʿAlī ibn Muḥammad al-Aṣbaḥī	II.5.1
YH*	MS Landberg-Brill (Leiden) 445 (now in Princeton?)	
	(1) *Mufīdat al-sāʾil ʿan ḥulul al-shams fi l-manāzil* (calendrical tables [?] for 1160 to 1253 Hijra) by Yaḥyā ibn Muḥsin ibn Aḥmad ibn Rājib	II.41.1(4)
	(2) Anonymous astronomical tables (four pages)	
	(3) *Kitāb al-Yawāqīt fī maʿrifat al-mawāqīt* (almanac) by Ḥusayn ibn Zayd ibn ʿAlī ibn Jaḥḥāf	I.39.1
	(4) Table showing configuration of the lunar mansions	
	(5) "Poem on the Syrian months and what one should eat and not eat during them," by ʿAbd Allāh ibn Asʿad al-Yāfiʿī	II.15.2
YI	MS Manchester John Rylands ar. 361A *Madkhal al-taʿlīm fī inshāʾ al-tasyīr wa-ʿamal al-taqwīm* by Abū Bakr ibn Abi l-Maʿalī	II.19.1
YJ	MSS Cairo DM (= Dār al-Kutub *mīqāt*) 949 and DJ (= *majāmīʿ* 705,7 and 709,23 (fols. 182r-184r) Three copies of *Manẓūma fi l-shuhūr al-Rūmīya* by ʿAbd Allāh ibn Asʿad al-Yāfiʿī	II.15.2
YK	MS Cairo DJ (= Dār al-Kutub *majāmīʿ*) 705,3 (fol. 19r) *Urjūza fi l-shuhūr al-Rūmīya* by Nishwān al-Ḥimyarī	II.3.2

YL	MS Cairo DJ (= Dār al-Kutub *mājamīʿ*) 713,14b (fol. 161v) Table for lunar crescent visibility by ʿAfīf al-Dīn (Ba-) Makhrama	II.26.2
YN	MS Cairo DM (= Dār al-Kutub *mīqāt*) 899 *Kitāb al-Shāmil fī dalāʾil al-qibla wa-ḥisāb al-Rūm wa-l-manāzil* attributed to the Abū Ḥamad Maḥfūẓ ibn ʿAbd al-Raḥmān al-Ḥaḍramī	II.26.1
YO	MSS Cairo TR (= Taymūr *riyāḍiyāt*) 322 and ṬJ (= Ṭalʿat *majāmīʿ*) 179,1 (fols. 1-79r) (Photographs of the Taymūr manuscript are preserved in the Dār al-Kutub, numbered K7194.) Two copies of *Sirāj al-tawḥīd al-bāhij al-nūr*. . . by ʿAbd Allāh ibn As adʿal-Yāfiʿī	II.15.1
YP*	MS Berlin Oct. 2542 *Ghāyat itqān al-ḥarakāt (zīj)* by ʿAbd Allāh ibn ʿAbd Allāh al-Sarḥī	II.37.1
YQ*	MS Princeton Garrett Hitti 971 *Nihāyat al-idrāk* by Muḥammad ibn Abī Bakr al-Fārisī	II.6.4
YR	MS Cairo DJ (= Dār al-Kutub *mājamīʿ*) 709,6 (fols. 29r-53v) [*Miftāḥ*] *al-asrār fī ʿilm al-falak al-dawwār* by Nūr al-Dīn ʿAlī ibn ʿAbd Allāh called at-Ṭawāshī (?)	II.27.1
YS*	MS Yale Nemoy 1476 (L-72a) Calendrical tables by Muḥammad ibn Aḥmad ibn al-Imām	II.41.1(12)
YT*	MS Princeton Garrett Hitti 1016 (61B) *Miftāḥ al-asrār fī ʿilm al-falak al-dawwār* by Nūr al-Dīn ʿAlī ibn ʿAbd Allāh al-Ṭawāshī (?)	II.27.1
YV	MS Aleppo Awqāf 968 Anonymous treatise *Kitāb al-Īḍāḥ al-shāfī* on folk astronomy	II.32.1
YW*	MS Berlin Glaser 210, fol. 164 (cf. *Ahlwardt*, no. 5871 on p. 270) *Qaṣīda* on the lunar mansions by al-Hādī ila l-ḥaqq	II.25.1
YX*	MS Vatidan V. 1139_3 *Qaṣīda* on the lunar mansions by al-Hādī ila l-ḥaqq	II.25.1
ZK	MS in the private collection of Sayyid Ahmad ʿAbd al-Qādir al-Ahdal, Zabid, Yemen. (A microfilm is available at the Arab League Institute of Arabic Manuscripts among the films prepared in the Yemen in the Spring of 1975.)	I.2

	(1)	*Zād al-musāfir* (*zīj*) by Muḥammad al-Daylamī	II.7.1; II.20.1; II.28.1
	(2)	Anonymous treatise on the horoscopes of Yemeni rulers	II.30.1
	(3)	Anonymous planetary tables for Sanaa	II.29.1
ZR*	(1)	MS Vatican V.1190,6	II.14.1
	(2)	MS Milan Ambrosiana CCCLV,N, fols. 188v-189r	II.14.1
	(3)	MS Rome Caetani 345, fol. 34r	II.14.1
		Qaṣīda on the lunar mansions by al-Hāmilī (three copies)	
	(4)	MS Milan Ambrosiana CCCLV,M, fols. 186v-188r	II.14.2
		Urjūza on the zodiac by al-Hāmilī	

LIST OF MANUSCRIPTS CONSULTED
(References are given in the list of sigla.)

Algiers Fagnan 1485,3 (AK)

Aleppo Awqaf 968 (YV)

Alexandria Municipal Library 3010D (AR); 5577C (AL)

Baghdad Awqāf 2982/6276 (AZ)

Berlin (= Deutsche Staatsbibliothek) Ahlwardt 5664 (BR); 5669 (BX); 5711 (BG); 5720 (BN); 5731 (BM); 5746 (BW); 5768 (BW); 5769 (BN); 5784 (BN); 5789 (BY); 5826 (BZ); 5883 (BG); 5888 (TJ); 5893 (BG); 5898 (BG); Oct. 2542 (YP). See also YW.

Cairo (= Dār al-Kutub al-Miṣrīya = Egyptian National Library)
ḥurūf M 73,2 (TK)
majāmīʿ 705,3 (YK); 705,7 (YJ); 709,6 (YR); 709,23 (YJ); 713,14b (YL)
mīqāt 145 (RA); 180 (TI); 191 (TI); 192 (TI); 400 (YA); 817 (TG); 899 (YN); 948 (TH); 949 (YJ); 983 (TI); 1015 (TL); 1196 (TI)
K (*falak wa-riyāḍa*) 4678 (SM); 7194 (YO); 7198 (TC); 16000 (SM)
Ṭalʿat *majāmīʿ* 179,1 (YO)
Ṭalʿat *mīqāt* 157,2 (TI); 248,1 (TI)
Taymūr *riyāḍiyāt* 105 (TA); 227,1 (TF); 274 (TC); 322 (YO)

Cambridge University Library Gg. 3.27 (CZ); Or. 1236 (YG)

Damascus Ẓāhirīya 3092 (DA)

Dublin Chester Beatty 4090 (DT); 4562 (DP)

Gotha Forschungsbibliothek A1523 (YB)

Hyderabad Asafiya (Andra Pradesh State Central Library) 395 (HD)

Hyderabad Salar Jung Museum astronomy 29 (HE)

Istanbul (Millet Genel) Ali Emiri 2722 (NP)

Istanbul Nurosmaniye 2951,1 (NE)

Istanbul (Suleymaniye) Hamidiye 830,2 (YC); Hosrev Paşa 216 (NQ)

Istanbul Topkapi H466 (YD)

Landberg-Brill (formerly in Leiden, now in Princeton?) 466 (YH); 78/141 (SR)

Leiden Universiteitsbibliotheek Or. 2807 (LQ)

London British Library Or. 3624 (LA); 3717 (LG); 3732 (LH); 3747 (LI); 3748 (LE); 3848 (LJ); 3849 (LF); 3906 (LK); 9116 (LM)

Manchester John Rylands ar. 361A (YI)

Milan Biblioteca Ambrosiana C46 (MA); C79 (MR); C83 (MU); C84 (MB); C86 (MV); C161 (MR); D21 (MR); D252 (MR); D268 (MR); D365 (MR); E16 (MT); E100 (MR); E170 (MR); E403 (MT); F67 (MR); F145 (MR); F201 (MT); F202 (MT); F267 (MT); Griffini 37 (ME). See also MF, MG, ZR

Oxford Bodleian Huntington 233 (ON); Marsh 134 (YF)

Paris Bibliothèque Nationale ar. 2523 (PA); 6840 (YE)

Princeton Garrett Hitti 971 (YQ); Hitti 1016 (YT); Mach 5015 (SQ)

Rome Caetani 345 (YR)

Sanaa Grand Mosque Library
falak 2 (SJ); 14 (SH); 101 (SN); 492 (SF)
majāmīʿ 23 (SM); 27 (SK)
unnumbered (SG, SI, SP)

Tehran Majlis al-Umma al-Īrānī (TZ)

Vatican ar. 949 (VK); 955 (VE); 962 (BV); 964 (VF); 1038 (VD); 1120 (VG). See also YR, YX, ZR

Yale Nemoy 1476 (YS)

Manuscripts in private collections in the Yemen:
Qadi Ismail al-Akwa (SA & SB)
Qadi Ali al-Sharafi (SC)
Qadi Muhammad al-Sharafi (SD)
Shaykh Ahmad al-Hatimi (SE)
Sayyid Ahmad al-Ahdal (ZK)
Sayyid Lutf Yusuf al-Mutawakkil (JA)
Qadi Husayn al-Sayyaghi (SL)

BIBLIOGRAPHICAL ABBREVIATIONS
Published Sources

Abu l-Fidāʾ M. Reinaud and MacG. de Slane, *Géographie d'Aboulféda: Texte Arabe*, Paris: Imprimerie Royale, 1840.

Ahlwardt W. Ahlwardt, *Die Handschriften-Verzeichnisse der Königlichen Bibliothek zu Berlin*, 17. Band: *Verzeichniss der arabischen Handschriften*, 5. Band, Berlin: A. Asher, 1893.

al- See name following al-

Anawati G. C. Anawati, "Trois talismans musulmans en arabe provenant du Mali (Marché de Mopti)," *Annales Islamologiques*, 11 (1972), pp. 287-339.

Azzawi A. El-Azzawi, *History of Astronomy in Iraq and its Relations with Islamic and Arab Countries in the Post Abbasid Periods* (in Arabic), Baghdad: Iraq Academy Press, 1959.

Banerjee & Sabra S. Banerjee and A. I. Sabra, "A Thirteenth-Century Magnetic Compass Described by Sultan al-Ashraf of Yemen," *Proceedings of the Second International Symposium on the History of Arabic Science*, Aleppo. 1979.

al-Bīrūnī 1 *Al-Qānūnuʾ l-Masʿūdī* (*Canon Masudicus*), 3 vols., Hyderabad: Osmania Oriental Publications, 1954-1956.

2 al-Bīrūnī, *Āfrād al-maqāl fī amr al-ẓilāl*, the second treatise in *Rasāʾiluʾl-Bīrunī*, Hyderabad: Osmania Oriental Publications, 1948.

Brockelmann C. Brockelmann, *Geschichte der arabischen Litteratur*, 2 vols., 2nd ed., Leiden: E. J. Brill, 1943-49, Supplementbände: 3 vols. Leiden: E. J. Brill, 1937-42.

Cairo Cat. D. A. King, *A Catalogue of the Scientific Manuscripts in the Egyptian National Library* (in Arabic), 2 vols., Cairo: General Egyptian Book Organization, in collaboration with the Smithsonian Institution and the American Research Center in Egypt, 1981-83 (?), and *A Survey of the Scientific Manuscripts in the Egyptian National Library* (in English) to be published by the American Research Center in Egypt with Undena Press.

Dorn B. Dorn, "Drei in der kaiserlichen öffentlichen Bibliothek befindliche astronomische Instrumente mit arabischen Inschriften," *Mémoires de l'Academie Imperiale des Sciences de St. Petersbourg*, 7e ser., 9:1 (1865).

Dozy — R. Dozy, *Supplément aux Dictionnaires Arabes*, 2nd ed., 2 vols., Leiden: E. J. Brill and Paris: Maisonneuve Frères, 1927.

Dozy & Pellat — *Le Calendrier de Cordoue*, ed. (R. Dozy) and tr. (C. Pellat), Leiden: E. J. Brill, 1961.

DSB — *Dictionary of Scientific Biography*, 15 vols. to date, New York: Charles Scribner's Sons, 1970 onwards.

EI_1 — *Encyclopedia of Islam*, 1st ed., 4 vols., Leiden: E. J. Brill, 1913-1934.

EI_2 — ____________________ , 2nd ed., 4 vols. to date, Leiden: E. J. Brill, 1960 onwards.

Ferrand — G. Ferrand, *Instructions Nautiques et Routiers Arabes et Portugais des XVe et XVIe Siècles*, Tôme III: *Introduction à l'Astronomie Nautique Arabe*, Paris: Paul Geuthner, 1928.

Gingerich & King — O. Gingerich and D. A. King, "Some Astronomical Observations from Thirteenth-Century Egypt," *Journal for the History of Astronomy*, 13 (1982), pp. 121-128.

Goldstein — B. R. Goldstein, "The Survival of Arabic Astronomy in Hebrew," *Journal for the History of Arabic Science*, 3 (1979), pp. 31-39.

Goldstein & Pingree — B. R. Goldstein and D. Pingree, "Astrological Almanacs from the Cairo Geniza," (in 2 pts.), *Journal of Near Eastern Studies*, 38 (1979), pp. 153-175 and 231-255.

Green & Stookey — A. H. Green and R. W. Stookey, "Research in Yemen: Facilities, Climate, and Current Projects," *Middle East Studies Association Bulletin*, 8 (1974), pp. 27-45.

Griffini 1 — E. Griffini, "Lista dei manoscritti arabi nuovo fondo della Biblioteca Ambrosiana," *RSO*, 3 (1910)–8 (1919-20). [See *Huisman*, p. 48 for details.]

2 — ____________________ , "Intorno alle stazione lunari," *RSO*, 1 (1907), pp. 423-438 and 607-608.

Gunther — R. T. Gunther, *The Astrolabes of the World*, 2 vols., Oxford: The University Press, 1932.

Ḥājjī Khalīfa — See *HK*.

al-Hamdānī — D. H. Muller, *Al-Hamdânî: Geographie der arabischen Halbinsel*, Amsterdam: Oriental Press, 1968 (Reprint).

Hawkins & King — G. S. Hawkins and D. A. King, "On the Orientation of the Kaʿba," *Journal for the History of Astronomy*, 13 (1982), pp. 102-109.

Henninger — J. Henninger, "Über Sternkunde und Sternkult in Nord- und Zentralarabien," *Zeitschrift für Ethnologie*, 79 (1954), pp. 82-117.

Hill — D. R. Hill, *The Book of Knowledge of Ingenious Mechanical Devices: Kitāb fī maʿrifat al-ḥiyal al-handasiyya by Ibn al-Razzāz al-Jazarī*, Dordrecht and Boston: D. Reidel, 1974.

HK — Ḥājjī Khalīfa, *Kashf al-ẓunūn an asāmi l-kutub wa-l-funūn*, 2 vols., Istanbul: Bahiya Press, 1941.

HS — *History of Science*.

Huisman — A. J. W. Huisman, *Les Manuscrits Arabes dans le Monde: une Bibliographie des Catalogues*, Leiden: E. J. Brill, 1967.

Ibn Abi l-Faḍāʾil — S. Kortentamer, *Aegypten und Syrien zwischen 1317 und 1341 in der Khronik des Mufaḍḍal b. Abi l-Faḍāʾil (fl. 1350)*, Freiburg: Klaus Schwartz Verlag, 1973.

Ibn Iyās — Muḥammad ibn Aḥmad Ibn Iyās, *Kitāb Taʾrīkh Miṣr al-mashhūr bi-Badāʾiʿ al-zuhūr fī waqāʾiʿ al-duhūr*, Bulaq, Cairo: Amiriya Press, 1311 Hijra (1893-4).

Janin & King 1 — L. Janin and D. A. King, "Ibn al-Shāṭir's *Ṣandūq al-Yawāqīt*: an astronomical 'compendium'," *Journal for the History of Arabic Science*, 1 (1977), pp. 187-256.

2 — ____________________, "Le Cadran Solaire de la Mosquée d'Ibn Ṭūlūn au Caire," *Journal for the History of Arabic Science*, 2 (1978), pp. 331-357.

Jazāʾirī — Ṭāhir al-Jazāʾirī, "al-Taʾlīf fi l-mulūk," *Majallat al-Majmaʿ al-ʿilmī al-ʿarabī bi-Dimashq* (*Journal of the Arab Academy in Damascus*), 27 (1952), pp. 52-60.

Kennedy 1 — E. S. Kennedy, "A Survey of Islamic Astronomical Tables," *Transactions of the American Philosophical Society*, N.S., 46:2 (1956), pp. 123-177.

2 — ____________________, *The Planetary Equatorium of Jamshīd Ghiyāth al-Dīn al-Kāshī (d. 1429)*, Princeton, N.J.: Princeton University Library, 1960.

Kennedy & Faris — N. Faris and E. S. Kennedy, "The Solar Eclipse Technique of Yaḥyā b. Abī Manṣūr," *Journal for the History of Astronomy*, 1 (1970), pp. 20-38.

al-Khazrajī — M. ʿAsal (ed.), *El-Khazreji's History of the Resúli Dynasty of Yemen*, 2 vols., and J. W. Redhouse (trans.), *The Pearlstrings: A History . . . by . . . El-Khazrejiyy*, 2 vols., Leiden: E. J. Brill and London: Luzac & Co., 1906-1918.

Khoury — R. G. Khoury, "Un fragment astrologique inédit attribué à Wahb b. Munabbih," *Arabica*, 19 (1972), pp. 139-144.

King 1 — D. A. King, "The Astronomical Works of Ibn Yunūs," Unpublished doctoral dissertation, Yale University, 1972.

2 — ____________________, "Ibn Yunūs' *Very Useful Tables* for Reckoning Time by the Sun," *Archive for History of Exact Sciences*, 10 (1973), pp. 342-394.

3 ______________________ , "A Double-Argument Table for the Lunar Equation Attributed to Ibn Yūnus," *Centaurus*, 18 (1974), pp. 129-146.

4 ______________________, "al-Khalīlī's *Qibla* Table," *Journal of Near Eastern Studies*, 34 (1975), pp. 81-121.

5 ______________________, "On the Astronomical Tables of the Islamic Middle Ages," *Studia Copernicana*, 13 (1975), pp. 37-56.

6 ______________________, "A Fourteenth-Century Tunisian Sundial for Regulating the Times of Muslim Prayer," in Y. Maeyama and W. G. Saltzer, eds., *Prismata: Festschrift für Willy Hartner*, Wiesbaden: Franz Steiner Verlag, 1977, pp. 187-202.

7 ______________________, "Astronomical Timekeeping in Fourteenth-Century Syria," *Proceedings of the First International Symposium for the History of Arabic Science*, (Aleppo, 1976), II, pp. 75-84.

8 ______________________, "Astronomical Timekeeping in Ottoman Turkey," *Proceedings of the International Symposium on the Observatories in Islam*, (Istanbul, 1977), pp. 245-269.

9 ______________________, "The Astronomy of the Mamluks," (Paper presented at the Symposium "Renaissance of Islam: Art of the Mamluks", Washington, D.C., 1981), to appear in *Muqarnas*.

10 ______________________, "Astronomical Alignments in Medieval Islamic Religious Architecture," *Annals of the New York Academy of Sciences*, 385 (1982), pp. 303-312.

11 ______________________, *The World About the Kaʿba: A Study of the Sacred Direction in Islam* (in preparation) to be published by Islamic Art Publications, S.p.A.

12 ______________________, "A Thirteenth-Century Yemeni Astrolabe in the Mestropolitan Museum of Art in New York," to appear.

Klein-Franke F. Klein-Franke, "A Hebrew Yemenite Manuscript of the Book "Elements of Astrology" by al-Bīrūnī," *Kiryat Sefer*, 47 (1972), p. 720.

Krause M. Krause, "Stambuler Handschriften islamischer Mathematiker," *Quellen und Studien zur Geschichte der Mathematik Astronomie und Physik*, B3:4 (1936), pp. 437-532.

Kunitzsch P. Kunitzsch, "On the Medieval Arabic Knowledge of Star Alpha Eridani," *Journal for the History of Arabic Science*, 1 (1977), pp. 263-267.

Lee S. Lee, "Notice of the Astronomical Tables of Mohammed Abibekr al-Farsi," *Transactions of the Cambridge Philosophical Society*, 1 (1822), pp. 249-265.

Levi della Vida — G. Levi della Vida, *Elenco dei Manoscritti Arabi Islamici della Biblioteca Vaticana* . . . , (Studi e Testi, no. 67), Vatican City: Biblioteca Apostolica Vaticana, 1935.

Löfgren & Traini — O. Löfgren and R. Traini, *Catalogue of the Arabic Manuscripts in the Biblioteca Ambrosiana*, Vol. I, Vicenza: Neri Pozza Editore, 1975.

MAES — D. A. King, *Mathematical Astronomy in Medieval Egypt and Syria* (in preparation).

al-Maqqārī — R. Dozy, G. Dugat, L. Krehl, and W. Wright, *Analectes sur l'Histoire et la Littérature des Arabes d'Espagne par al-Makkārī*, 2 vols., Leiden: E. J. Brill, 1855-1860 and 1858-1861.

Mayer — L. A. Mayer, *Islamic Astrolabists and Their Works*, Geneva: Ernst Kundig, 1956.

Nallino — C. A. Nallino, *Raccolta di Scritti Editi e Inediti*, vol. V: Astrologia–Astronomia–Geografia, Rome: Instituto per l'Oriente, 1944.

Neugebauer — O. Neugebauer, "Studies in Byzantine Astronomical Terminology," *Transactions of the American Philosophical Society*, 50:2 (1960).

Pingree 1 — D. Pingree, "Gregory Chioniades and Palaeologan Astronomy," *Dumbarton Oaks Papers*, 18 (1974), pp. 135-160.

2 — ______________________, "The Greek Influence on Early Islamic Mathematical Astronomy," *Journal of the American Oriental Society*, 93 (1973), pp. 32-43.

Price — D. J. de S. Price, "Gears from the Greeks: the Antikythera Mechanism–a Calendar Computer from ca. 80 B.C.", *Transactions of the American Philosophical Society*, 64:7 (1974) (also published New York: Science History Publications, 1975).

al-Qifṭī — J. Lippert, (ed.), *Ibn al-Qifṭī's Taʾrīḫ al-Ḥukamā* , Leipzig: Dieterich'sche Verlag, 1903.

Reinhart — A. K. Reinhart, "Manuscript Research in the Yemen (Arab Republic)," *Middle East Studies Association Bulletin*, XIV:2 (December 1980), pp. 22-30.

RSO — *Rivista degli Studi Orientali*.

SATMI — D. A. King, *Studies in Astronomical Timekeeping in Medieval Islam*. I: A Survey of Medieval Islamic Tables for Reckoning Time by the Sun and Stars; II: A Survey of Medieval Islamic Tables for Regulating the Times of Prayer; III: A Survey of Medieval Islamic Shadow Schemes for Simple Time-Reckoning; IV: On the Origin of the Prayers in Islam; and V: On the Role of the Muwaqqit in Medieval Islamic Society; to appear.

Sayili — A. Sayili, *The Observatory in Islam*, Ankara: Turkish Historical Society (Series VII, No. 38), 1960.

Sayyid — A. F. Sayyid, *Sources de l'Histoire du Yémen à l'Epoque Musulmane*, Cairo: Institut Français d'Archéologie Orientale du Caire, 1974.

Sédillot-fils — L. A. Sédillot, "Mémoire sur les instruments astronomiques des arabes," *Mémoires presentés . . . à l'Academie Royale . . . de l'Institut de France*, 1. sér., vol. 1 (1844).

Sédillot-père — J.-J. Sédillot, *Traité des Instruments Astronomiques des Arabes Composé au Treizième Siècle par Aboul Hhassan Ali de Maroc Intitulé Jāmiʿ al-mabādiʾ wa-l-ghāyāt*, 2 vols., Paris: Imprimerie Royale, 1834-35.

Serjeant 1 — R. B. Serjeant, "Star-Calendars and an Almanac from South-West Arabia," *Anthropos*, 49 (1954), pp. 433-459.

2 — ______, *The Portuguese off the South Arabian Coast*, Beirut: Libraire du Liban, 1974 (first published by Oxford University Press, 1963).

Sezgin — F. Sezgin, *Geschichte des arabischen Schrifttums*, 7 vols. to date, Leiden: E. J. Brill, 1967 onwards.

Savage-Smith & Smith — E. Savage-Smith and M. B. Smith, *Islamic Geomancy and a Thirteenth-Century Divinatory Device*, Malibu, Ca.: Undena Publications, 1980.

Storey — C. A. Storey, *Persian Literature: a Bio-Bibliographical Survey*, Vol. II, London: Luzac and Co., 1958.

Suter — H. Suter, "Die Mathematiker und Astronomen der Araber und ihre Werke," *Abhandlungen zur Geschichte der mathematischen Wissenschaften*, 10 (1900) (quoted by number), and "Nachträge und Berechtigungen zu "Die Mathematiker und Astronomen der Araber und ihre Werke," ibid., 14 (1902), pp. 157-185 (quoted by number and the abbreviation N).

Traini — R. Traini, "Les Manuscripts Yéménites dans les Bibliothèques d'Istanbul," *Revue d'Histoire des Textes*, 3 (1973), pp. 203-230.

Ullmann 1 — M. Ullmann, *Die Medizin im Islam*, (Handbuch der Orientalistik, 1. Abt., Band VI:I), Leiden: E. J. Brill, 1970.

2 — ______, *Die Natur- und Geheimwissenschaften im Islam*, (Handbuch der Orientalistik, 1. Abt., Band VI:2.), Leiden: E. J. Brill, 1972.

al-Wāsiʿī — ʿAbd al-Wāsiʿ ibn Yaḥyā al-Wāsiʿī, *Kanz al-thiqāt fī ʿilm al-awqāt*, Cairo: Ḥijāzī Press, 1359 Hijra (= 1939).

Wiedemann — E. Wiedemann, "Ein Instrument, das die Bewegung von Sonne und Mond darstellt, nach al-Bīrūnī," *Der Islam*, 4 (1913), pp. 5-13.

INDEX

Index of Personal Names

In the alphabetical arrangement of this index the words al- (the), and ibn (son of) have been ignored, although not when the name begins with Ibn. No references to Section 1.6 are included here. The sections in which the authors are the main subject are printed in boldface type.

ʿAbd Allāh ibn ʿAbd Allāh, see al-Muthannā al-Sarḥī
ʿAbd Allāh ibn ʿAbd al-Raḥmān, **II.41.1**; App. B, *ad* p. 484
ʿAbd Allāh ibn Aḥmad al-Shamākhī (?), App. B, *ad* p. 488
ʿAbd Allāh ibn Aḥmad al-Shatmakhī (?), **II.41.1**
ʿAbd Allāh ibn ʿAlī ibn al-ʿAbbās (*imām* of Yemen), I, note 16
ʿAbd Allāh ibn Asʿad, see al-Yāfiʿī
ʿAbd Allāh ibn Ḥamza al-Dawwārī, **II.41.1**; App. B, *ad* p. 488; Addenda, no. 1
ʿAbd Allāh ibn Muḥammad, **App. A, 11**
ʿAbd Allāh ibn Salāḥ ʿAnqūb, App. B, *ad* p. 486
ʿAbd Allāh ibn Ṣalāḥ Dāʿir, **II.3.1**
ʿAbd Allāh al-Sarḥī, see al-Muthannā al-Sarḥī
ʿAbd Allāh ibn ʿUmar, see Bā Makhrama
ʿAbd Allāh ibn ʿUmar ibn Khalīl, **App. A, 8**
ʿAbd al-Laṭīf ibn Aḥmad, App. A, 2.1
ʿAbd al-Qādir ibn Aḥmad, App. B, *ad* p. 494
ʿAbd al-Qayyūm al-Raghīlī, App. B, *ad* p. 485
ʿAbd al-Rāziq (?), II.2.1
ʿAbd al-Razzāq (?), II.2.1
ʿAbd al-Wahhāb ibn ʿAlī, **II.41.1**
ʿAbd al-Wāsiʿ ibn Yaḥyā, see al-Wāsiʿī
al-ʿAbdī, see Isḥāq ibn Muḥammad
Abraham, II.6.7
Abū l-ʿAbbās ibn al-Qāṣṣ, see Ibn al-Qāṣṣ
Abū ʿAlī ʿArafa, see ʿArafa
Abū Bakr ibn Abi l-Maʿālī, see Ibn Abi l-Maʿālī
Abū Bakr ibn ʿAlī, see al-Hāmilī
Abū Bakr ibn Ismāʿīl, see Ibn al-Mushrif (Egyptian astronomer)
Abū Bakr ibn Muḥammad al-Fārisī, II.6, note 3
Abu l-Fatḥ al-Daylamī, see al-Nāṣir
Abu l-Fidāʾ (Syrian scholar/prince), II.9.0; II.9, note 2
Abū Jaʿfar al-Baṣrī, II.5.1
Abū Jaʿfar al-Rāsibī, II.2.1
Abu Jaʿfar ʿUmar (Rasulid *wazīr* and *qāḍī*), II.13.1
Abu l-Qāsim al-Balkhī (Transoxanian astronomer), I, note 16
Abu l-Wāsim al-Ḥyhy (?) al-Makkī, **II.42**
Abu l-Ṣalt Umayya ibn Abi l-Ṣalt (Andalusian astronomer), I.4; II.8.2
Abu l-ʿUqūl, I.3; I.4; I, note 10; **II.9**; II.10.0; II.12.1; II.18.1; II.28.1; II.38.0
al-Afḍal (Rasulid Sultan), I.3; I.4; II.9.0; II.9.3; II.10.1; **II.18**
al-Ahdal, see Muhammad ibn ʿAbd al-Raḥmān
Aḥmad ibn Abī Aḥmad, see Ibn al-Qāṣṣ
Aḥmad al-Ḥusaynī, **App. A, 10**
Aḥmad ibn Ibrahīm, see al-Ashʿarī
Aḥmad al-Mʿd (?) al-Rāzī (?) (muezzin in Mecca), II.2.1
Aḥmad ibn Muḥammad, see al-Ashʿarī, al-Bakhāniqī, Ibn al-Hāʾim
Aḥmad ibn Muḥammad al-Dawwārī, Addenda, no. 3
Aḥmad ibn Muḥammad al-Khālidī, Addenda, no. 3
Aḥmad ibn Mūsā al-Jallād, **App. A, 5**
Aḥmad ibn Mūsā al-ʿUjayl, App. A, 5.0
Aḥmad ibn Ṣalāḥ, II.2.1
Aḥmad ibn ʿUmar, see al-Murshidī, al-Khuzāʿī
Aḥmad ibn Yaḥyā al-Dawwārī, **II.34**
Aḥmad ibn Yaḥyā al-Mahdī, **II.24a**
al-Akwaʿ, see ʿAlī ibn Ḥasan
Alexander, II.6.7
ʿAlī ibn ʿAbd Allāh, see al-Ṭawāshī
ʿAlī ibn ʿAbd Allāh al-Zilʿī, App. A, 5.0
ʿAlī ibn ʿAbd Allāh al-ʿAydarus, II.41.1
ʿAlī ibn Aḥmad al-Jallād, **App. A, 6**
ʿAlī ibn ʿAlī al-Yamanī, II.44.1

ᶜAlī ibn Ḥasan al-Akwaᶜ, II.41.1; App. B, *ad* p. 487
ᶜAlī ibn Ibrāhīm, see Ibn al-Shāṭir
ᶜAlī ibn ᶜĪsā (Abbasid astronomer), II.6, note 12
ᶜAlī ibn Mankabris (visitor to Yemen), I.5
ᶜAlī ibn Muḥammad, **II.41.1**
al-ᶜĀmirī, see Muḥammad ibn Surāqa
Anonymous, II.11; II.16; II.17; II.21; II.29; II.30; II.32; II.35; II.36.1; II.40; II.41.1; II.48; App. A, 12
ᶜAnqūb, see ᶜAbd Allāh ibn Ṣalāḥ
al-ᶜAnsī, **II.46**
ᶜArafa (muezzin in Fustat), II.2.1
Ardrūmī, see Muḥammad ibn Aḥmad
al-Aṣbahī, I.3; I, note 13; **II.5**; II.32.1; App. B, *ad* p. 483
al-Ashᶜarī, App. A, 2
al-Ashraf (Rasulid Sultan), I.3; I.4; I.5; I, notes 9, 10, and 13; II.6.6; **II.8**; II.10.0
al-Ashraf II (Rasulid Sultan), II.18a; II.19.1; App. A, 5.0
al-ᶜAṭṭās, see Muhammad al-ᶜAṭṭās
al-ᶜAydarus, see Muḥammad al-ᶜAydarus

Ba ᶜAyyād, see Maḥẓūẓ ibn ᶜAbd al-Raḥmān
Bā Bashīr, **II.41.1**
Bā Faḍl, see ᶜAbd Allāh ibn ᶜAbd al-Raḥmān
Ba Makhrama, **II.26**; App. B, *ad* pp. 484-485
Ba Ṣabrayn, see al-Shibāmī
Badr al-Dīn Muḥammad ibn Abī Bakr al-Fārisī, see II.6, notes 1 and 3
Bāghūth al-Ḥaḍramī, App. B, *ad* p. 486
Bahram, II.6.7
al-Bakhāniqī (Egyptian astronomer), I.4; **II.13**
al-Balkhī, see Abu l-Qāsim
al-Baṣrī, see Abū Jaᶜfar
al-Battānī (Syrian astronomer), II.24.0; II.24, note 2
al-Bawsī, Addenda, no. 3
al-Bīrūnī (Transoxanian astronomer), I.3; I, notes 16 and 17; II.6, note 12
Brahmans, II.6.7

Dāʾūd, son of al-Malik Manṣūr (Rasulid ruler of Yemen), II.6, note 3
al-Dawwārī, see ᶜAbd Allāh ibn Ḥamza, Aḥmad ibn Yaḥyā, Luṭf Allāh ibn ᶜAbd Allāh
al-Daylamī (astronomer), I.3; I.4; I, note 13; II.7.1; II.20.1; II.23.1, II.28; II.29.1
al-Daylamī (*imām*), see al-Nāṣir
al-Dimashqī, see ᶜAbd al-Laṭīf ibn Aḥmad
Dorotheus, II.8.1

al-Faḍl ibn Abī Saᶜd, see al-Uṣayfirī
al-Fahhād, I.4; II.6.3
al-Farghānī, I.4; I.6; II.8.2; II.13.1
al-Fārisī, I.3; I.4; I, notes 9, 10, and 16; II.5a.0; **II.6**; II.7.0; II.21.1; II.42.1; App. B, *ad* p. 483
al-Fihrī, see Ḥasan ibn ᶜAlī
al-Fīrūzābādī, II.32.1

al-Ghazālī, II.42.1
al-Gīlī, see Kūshyār

al-Hādī, see ᶜIzz al-Dīn
al-Hādī ibn ᶜAlī, App. B, introd.; App. B, *ad* p. 487
al-Hādī ila l-ḥaqq, see ᶜIzz al-Dīn
al-Ḥaḍramī, see Bāghūth, Maḥfūẓ ibn ᶜAbd al-Raḥmān, Maḥẓūẓ ibn ᶜAbd al-Raḥmān
Ḥājjī Khalīfa (Turkish bibliographer), I.3; II.5a.0; II.6, note 3; II.6.5; I.7.1; II.8, note 2; App. A, 1.0; App. A, 1.2; App. A, 2.0; App. A, 4.1
al-Ḥākim (Fatimid Caliph), I.3; II.8.2
al-Ḥalabī, see Yaḥyā ibn Taqi l-Dīn
al-Hamdānī, I.3; I.4; I, notes 9 and 10; **II.1**
al-Hāmilī, **II.14**; App. A, 1.1; **App. A, 4**
al-Ḥarīrī (?), see II.18.1
al-Ḥarrānī, see Sinān ibn al-Fatḥ
al-Ḥasan ibn ᶜAbd Allah, see al-Sarḥī (I)
al-Ḥasan ibn Aḥmad al-Hamdānī, see al-Hamdānī
al-Ḥasan ibn ᶜAlī, see al-Marrākushī
Ḥasan ibn ᶜAlī al-Fihrī, II.8.2
al-Ḥasan ibn Ḥumayd, App. B, *ad* p. 484
al-Ḥasan al-Sarḥī, see al-Sarḥī
Ḥasan ibn Shams al-Dīn al-Jaḥḥāf, see Ibn Jaḥḥāf
Ḥasan ibn Zayd, see Ibn Jaḥḥāf
al-Ḥaṭṭāb (?), I, note 16
al-Ḥimyarī, see Nashwān ibn Saᶜīd
Ḥulagu, II.8.2
Ḥusayn, al-Manṣūr Abu l-ᶜAbbās, II.38.1
Ḥusayn ibn Zayd, see Ibn Jaḥḥāf
al-Ḥusaynī, see Aḥmad al-Ḥusaynī, Muḥammad ibn Aḥmad al-Ḥusaynī
al-Ḥusnī, see Muḥammad ibn Aḥmad al-Ḥusaynī
al-Ḥyhy (??), see Abu l-Qāsim

al-Ibarī, see Muḥammad ibn Abī Bakr al-Ibarī
Ibn Abi l-Faḍāʾil (Egyptian historian), I.5; II.8, note 8
Ibn Abi l-Maᶜālī, I.3; I, note 10; **II.19**
Ibn Abi Rijāl, II.23.0
Ibn Abi l-Ṣalt (Andalusian astronomer), I.4; II.8.2
Ibn al-ᶜAnz, see Muḥammad ibn Aḥmad ibn ᶜIzz al-Dīn
Ibn ᶜArīb, II.4.1
Ibn al-Dahhān (Syrian astronomer), II.18.1
Ibn Dāᶜir, see ᶜAbd Allāh ibn Ṣalāḥ
Ibn al-Ḥāʾik, see al-Hamdānī
Ibn al-Hāʾim, App. A, 3.1

Ibn al-Ḥājib, II.6, note 3
Ibn Hāshim, II.6.7
Ibn Isḥāq (Maghribi astronomer), I.3; I.4; II.28.1; II.28, note 2
Ibn Iyās (Egyptian historian), I.5
Ibn al-Jaḥḥāf (all references to same individual??), I, note 10; II.38.0; **II.39**; **II.41.1**; **App. A, 9**; App. B, *ad* p. 487
Ibn al-Mushrif, **II.12**; II.18.1
Ibn al-Mushrif (Egyptian astronomer), II.12, note 1
Ibn Nawbakht (Abbasid astrologer), II.8.1
Ibn al-Qāṣṣ (tenth-century legal scholar from Tabaristan), II.5.1
Ibn al-Qifṭī, see al-Qifṭī
Ibn Raḥīq, I.3; **II.2**; II.6.1
Ibn Rājib, I, note 10; **II.41.1**; Addenda, no. 4
Ibn Salm, see Muḥammad ibn ʿAbd Allāh
Ibn al-Shāṭir (Syrian astronomer), App. B, introd.; App. B *ad* p. 485
Ibn Surāqa, see Muḥammad ibn Surāqa
Ibn Yūnus (Egyptian astronomer), I.3; I.4; I.5; I, note 7; II.7.1; II.8.2; II.9.1; II.9.3; II.9, note 4; II.11.1; II.17.1; II.23.1; II.24, note 2; II.28.1; II.29.1; II.31.1
Ibrāhīm ibn ʿAlī, see al-Aṣbaḥī
Ibrāhīm ibn al-Mahdī, App. A, 9.1
Ibrāhīm ibn Mamdūd al-Ḥāsib, II.8.2
Ibrāhīm ibn Muḥammad, see al-Bawsī
Ibrāhīm ibn Yaḥyā al-ʿIlfī, **II.41.1**
Ibrāhīm ibn Zayd, see Ibn Jaḥḥāf
al-Ibrī, see al-Ibarī
al-ʿIlfi, see Ibrāhīm ibn Yaḥyā
Imām al-Mahdī, II.44.1
al-Iṣfahānī, see Muḥammad ibn Abī Bakr al-Iṣfahānī
Iṣhāq ibn Muḥammad, App. B, *ad* pp. 486-487
Iṣhāq ibn Yūsuf, see al-Ṣardafī
Ismāʿīl ibn al-ʿAbbās, see al-Ashraf II
Ismāʿīl ibn Aḥmad al-Najrānī, App. B, *ad* p. 484
Ismāʿīl ibn ʿAṭīya al-Najrānī, I.4; **II.23**; II.24.0; II.28.1; App. B, *ad* p. 484
ʿIzz al-Dīn ibn al-Ḥusayn, **II.25**; App. B, *ad* p. 485

al-Jaghmīnī (Transoxanian astronomer), Addenda, no. 6
al-Jaḥḥāf, see Ibn Jaḥḥāf
al-Jallād, see ʿAlī ibn Aḥmad, Aḥmad ibn Mūsā
Jāmasp, II.6.7
al-Janadī, see al-Aṣbaḥī
al-Jazarī (Syrian author on automata), I, note 22
Jesus Christ, II.6.7
al-Jīlī, see Kūshyār

al-Kaʿbī, II.20
Kalārjī, see al-Maḥallī
al-Kāmil (Mamluk Sultan), I.5
al-Kāshī (Transoxanian astronomer), II.6, note 12
Kātib Chelebī, see Ḥājjī Khalīfa
al-Kawāshī, I.3; I.4; I, note 10; II.6, note 3; **II.7**; II.28.1; II.38.0
al-Kawkabānī, see ʿAbd al-Qādir ibn Aḥmad
al-Kawwās, II.7, note 1
al-Kawwāsh, II.7, note 1
al-Khālidī, see Aḥmad ibn Muḥammad al-Khālidī
al-Kharaqī (Khurasanian astronomer), I, note 16
al-Khāṣṣ, see al-Ṣiddīq ibn Muḥammad
al-Khaṭṭāb (?), see al-Ḥattāb (?)
al-Khazrajī (Yemeni historian), II.6.0; II.6.5
al-Khuzāʿī, App. A, introd.; **App. A, 3**
al-Khwārizmī (Abbasid astronomer), I.6E; II.8.2; App. A, introd.; App. A, 3.1
Kūshyār ibn Labbān (Buwayhid astronomer), I, note 16; II.6.1; II.8.1

Luṭf Allāh ibn ʿAbd Allāh, App. B, *ad* p. 489

al-Madhḥijī (?), see ʿAbd Allāh ibn Muḥammad
al-Madhjīḥī (?), see Aḥmad ibn ʿUmar
al-Maghribī, see Ibn Isḥāq
al-Maḥallī, I, notes 10 and 16; II.7.1; II.9.3; II.35.1; **II.38**
al-Mahdī, II.44.1
al-Mahdī Aḥmad ibn Yaḥyā, see Aḥmad ibn Yaḥyā
Maḥfūẓ (?) ibn ʿAbd al-Raḥmān, II.26.1; App. B, *ad* p. 486
Maḥmūd ibn Muḥammad, see al-Jaghmīnī
Maḥẓūẓ (?) ibn ʿAbd al-Raḥmān, II.26.1; App. B, *ad* p. 486
al-Makkī, see Abu l-Qāsim
al-Maʾmūn (Abbasid Caliph), II.6.1
Mani, II.6.7
al-Manṣūr (Rasulid Sultan), see Dāʾūd
al-Manṣūr Abu l-ʿAbbās Ḥusayn (Yemeni Sultan), II.38.1
al-Manṣur bi-llāh al-Qāsim ibn Muḥammad (Yemeni Sultan), II.41.1
al-Maqqārī (Algerian historian), I.5; I, note 9
al-Maqrīzī (Egyptian historian), I, note 28
al-Marrākushī (Cairene astronomer), I.4; I, note 16; II.6, note 12; II.8.2; II.18.1
Māshāʾallāh (Abbasid astrologer), I, note 16
Maʿshūq al-Qararī, II.2.1
al-Masʿūd (Ayyubid ruler of Yemen), II.4.0
Mawdūd ibn ʿUthmān al-Shirwānī, I, note 16
Mazdak, II.6.7
al-Maziḥfī, see al-Khuzāʿī
al-Miqrāʾī, see al-Ḥasan ibn Ḥumayd
Moses, II.6.7

al-Mu'ayyad (Rasulid Sultan), I.4; II.9.0; **II.10**; II.11.1; II.18.1
Muḥammad (the Prophet), II.6.7
Muḥammad ibn ʿAbd Allāh . . . ibn Salm, App. A, 1.1; **App. A, 7**
Muḥammad ibn ʿAbd al-Laṭīf, see al-Thābitī
Muḥammad ibn ʿAbd al-Raḥmān, App. B, *ad* p. 488
Muḥammad ibn Abī Bakr, see al-Fārisī, al-Kawāshī, al-Shillī
Muḥammad ibn Abī Bakr al-Rashīdī al-Ibarī al-Iṣfahānī, II.6.5
Muḥammad ibn Aḥmad, see Abu l-ʿUqūl
Muḥammad ibn Aḥmad Āghā Ardrūmī, II.38.0
Muḥammad ibn Aḥmad al-Ḥusaynī, App. B, *ad* pp. 487-488
Muḥammad ibn Aḥmad ibn al-Imām, **II.41.1**
Muḥammad ibn Aḥmad ibn ʿIzz al-Dīn, App. B, *ad* p. 485
Muḥammad ibn Aḥmad al-Khuzāʿī, see al-Khuzāʿī
Muḥammad ibn Aḥmad al-Najjār, II.41.1
Muḥammad ibn ʿAlī, see al-Daylamī
Muḥammad al-ʿAṭṭās ibn Salām, II.41.1
Muhammad al-ʿAydarus ibn ʿAbd Allāh, **II.41.1**
Muḥammad al-Ḍaʿīf al-Saqqāf, II.41.1
Muḥammad ibn Ḥāmid, App. B, *ad* p. 488
Muḥammad Ḥaydara, **II.44**
Muḥammad ibn Raḥīq, see Ibn Raḥīq
Muḥammad ibn al-Ṣiddīq al-Khāṣṣ, App. B, *ad* p. 485
Muḥammad ibn Surāqa al-ʿĀmirī, II.2.1
Muḥammad ibn ʿUmar, App. B, *ad* p. 484
Muḥammad ibn Yaḥyā ibn Abī Manṣūr (Abbasid astronomer), II.6.1
al-Mujāhid (Rasulid Sultan), I, note 10; **II.11a**; II.13.1; II.18.0; II.18a
Muslim ibn Maḥmūd, see al-Shayzarī
al-Muthannā al-Sarḥī, I.3; I.4; I, notes 10 and 20; II.36.0; **II.37**; II.41.1
al-Muẓaffar (Rasulid Sultan), I.3; I.4; I.5; **II.5a**; II.6.0; II.6.2; II.6.3; II.6.4; II.7.1; II.10.0

al-Najrānī, see ʿAbd Allāh ibn Muḥammad, Ismāʿīl ibn Aḥmad, Ismāʿīl ibn ʿAṭīya, Zayd ibn ʿAṭīya
Nashwān ibn Saʿīd al-Ḥimyarī, **II.3**
al-Nāṣir (Mamluk Sultan), I.5
al-Nāṣir (Rasulid Sultan), II.22.1
al-Nāṣir Abu l-Fatḥ al-Daylamī (eleventh-century *imām*), II.28.0
Nāṣir al-Dīn al-Ṭūsī, see al-Ṭūsī
Naṣr ibn Naṣr, II.6, note 3
al-Nassāba, see al-Ashʿarī
Noah, II.6.7
al-Nuʿmānī, App. B, *ad* p. 483
Nur al-Dīn al-Ṭawāshī, see al-Ṭawāshī

Ptolemy, II.1.2; II.6.2; II.29.1

al-Qararī, see Maʿshūq
al-Qāsim ibn Muḥammad, see al-Manṣūr
al-Qifṭī (Syrian historian of science), I, note 9; II.1.0; II.1.1; II.1.5

al-Raghīlī, see ʿAbd al-Qayyūm
al-Rashīdī, see Muḥammad ibn Abī Bakr al-Rashīdī
al-Rāsibī, see Abū Jaʿfar
Riḍwān ibn ʿAbd Allāh, see Riḍwān Efendī
Riḍwān Efendī (Egyptian astronomer), II.38.0

al-Ṣaʿdī, see al-Dawwārī
Sahl ibn Bishr (Abbasid astrologer), I, note 16
Saʿīd ibn ʿUmar, **II.41.1**
Ṣalāḥ al-Dīn Yūsuf, see al-Masʿūd
Samaw'al ibn Yaḥyā al-Maghribī (astronomer in Iran), App. A, introd.
al-Saqqāf, see Muḥammad al-Ḍaʿīf, Muḥammad ibn Ḥāmid, ʿUmar ibn Saqqāf
al-Ṣardafī, **App. A, 1**; App. A, 4.1; App. A, 7.1
al-Sarḥī (ʿAbd Allāh ibn ʿAbd Allāh), see al-Muthannā al-Sarḥī
al-Sarḥī (al-Ḥasan ibn ʿAbd Allāh), **II.36**; II.37.0; II.37.1
al-Sarmī, see (al-)Hādī ibn ʿAlī
Sayf al-Dīn Abu l-Hasan ʿAlī ibn Mankabris, see ʿAlī ibn Mankabris
al-Ṣaymarī (Abbasid astrologer), I, note 16
al-Shajarī, see Aḥmad ibn Muḥammad
al-Shamākhī (?), see ʿAbd Allāh ibn Aḥmad
al-Shatmakhī (?), see ʿAbd Allāh ibn Aḥmad
al-Shawqānī, **II.45**
al-Shayzarī, **II.4**; Addenda, no. 8
al-Shibāmī, II.33.1; **II.43**
al-Shillī, App. B, *ad* pp. 485-486
al-Shirwānī, see Mawdūd ibn ʿUthmān
al-Ṣiddīq ibn Muḥammad al-Khāṣṣ, II.41.1
Sinān ibn al-Fatḥ al-Harrānī, II.8.1
al-Suyūṭī (Egyptian polymath), I, note 16

al-Ṭawāshī, **II.27**
al-Thābitī, **II.33**; II.43.1
al-Tujībī (Andalusian astronomer), I.5
al-Tūnisī, see Ibn Isḥāq
al-Ṭūsī (Persian scientist), I.5; II.8.2; II.17.1

al-ʿUjayl, see Aḥmad ibn Mūsā
Ulugh Beg (Transoxanian scholar/prince), I.4; II.38.1
ʿUmar ibn ʿAbd Allāh, **II.41.1**
ʿUmar ibn Saqqāf, App. B, *ad* p. 487
ʿUmar ibn Yūsuf, see al-Ashraf
Umayya ibn Abi l-Ṣalt, see Abu l-Ṣalt
al-ʿUṣayfirī, Addenda, no. 3

Wahb ibn Munabbih, I, note 30
al-Wāsiʿī, **II.41.1**; **II.47**; II.48.1
al-Wathīq bi-llāhi, II.28.0

al-Yāfiʿī, **II.15**; II.26.1; II.42.1
Yaḥyā (*imām* of Yemen), II.47.2
Yaḥyā ibn Abī Manṣūr (Abbasid astronomer), II.6.1; see also *Mumtaḥan Zīj*
Yaḥyā ibn Maẓhar, App. B, *ad* p. 488
Yaḥyā ibn Muḥammad al-Ḥaṭṭāb (?), I, note 16
Yaḥyā ibn Muḥsin, see Ibn Rājib
Yaḥyā ibn Muṭahhar (?), see Yaḥyā ibn Maẓhar
Yaḥyā ibn Taqi l-Dīn al-Ḥalabī (Syrian scholar), App. A, 2.1
Yūsuf ibn ʿAbd Allāh, II.38, note 1
Yūsuf, al-Malik al-Masʿūd, see al-Masʿūd
Yūsuf Kalārjī, see al-Maḥallī
Yūsuf ibn Yūsuf al-Maḥallī, see al-Maḥallī

Zacuto (Andalusian astronomer), I, note 16; II.38.0
Zarathustra, II.6.7
Zayd ibn ʿAṭīya al-Najrānī, **II.24**
al-Zilʿī, see ʿAlī ibn ʿAbd Allāh

Index of Titles

In the alphabetical arrangement of this index the abbreviations *K.* = *Kitāb* (book), *R.* = *Risāla* (treatise), and also *fī* (concerning), *al-* (the), *etc.*, are ignored. No references to Section I.6 are included here. Dubious titles, mainly from Appendix B, are indicated with a question mark. The sections in which the works are discussed in detail are printed in boldface type.

ʿAdāt al-nujūm, **II.4.1**
ʿAlāʾī Zīj, see *al-Zīj al-ʿAlāʾī*
Almagest, II.1.2
Almanac, anonymous, Addenda, nos. 2 and 7
Almanac of Ibn al-Jaḥḥāf, II.38.0
Almanac for Sanaa, I.3; I.4; I, note 28; **II.11.1**
Almanac for Taiz, I.3; I.4; I, note 28; **II.22.1**
"Almanac" of Zacuto, I, note 16; II.38.0
R. fi l-ʿAmal . . . ʿala (?) ruʾyat al-hilāl, App. B, *ad* p. 494
R. fī ʿAmal al-asṭurlāb, II.10.1
R. fi l-Asṭurlāb, App. B, *ad* pp. 485-486
R. Awāʾil al-shuhūr (?), App. B, *ad* pp. 486-487

al-Badr al-ṭāliʿ, II.45.1
Balīgh Zīj, see *al-Zīj al-Balīgh*
R. fī Bayān ḍābiṭat ʿuqūd al-aʿdād, **App. A, 10.1**
Bughyat al-ṭālib, see *Zīj* of al-Sarḥī
Bulghat al-muqtāt, **II.41.1**; Addenda, no. 1

Cairo corpus of tables for timekeeping, I.4; II.9.3; II.13, note 1
Calendar of Cordova, II.4.1
Compendium of al-Afḍal, I.3; I.4; II.9.3; **II.18.1**
Corpus of tables for timekeeping, see Cairo, Damascus, Taiz

al-Dalāʾil fī aḥkām al-nujūm, App. B, *ad* p. 485
K. Dalāʾil al-qibla (Ibn al-Qāṣṣ), II.5.1
K. Dalāʾil al-qibla (Ibn Surāqa), II.2.1
Damascus corpus of tables for timekeeping, II.9.3
K. al-Ḍarb al-Hindī, App. A, 1.1
al-Durr al-naẓīm, App. B, *ad* p. 485
K. al-Durra al-muntakhaba, II.6, note 3

K. fi l-Falak (?), App. B, *ad* p. 486
Fawāʾid fī maʿrifat al-aẓlāl, **II.26.3**
R. al-Fawāʾid wa-l-asrār, App. B, *ad* p. 484

R. fi l-Ghālib wa-l-maghlūb, II.1, note 1
Ghāyat itqān al-ḥarakāt, see *Zīj* of al-Muthannā al-Sarḥī

Ḥākimī Zīj, see *al-Zīj al-Ḥākimī*
al-Hayʾa al-Saniya. . . , I, note 16
al-Hindī, see al-Ṣardafī (author)
al-Hindī fī ʿilm al-farāʾiḍ, App. A, 1.3
Ḥisāb al-Shibāmī, II.33.1; **II.43.1**

K. Ibn al-Mushrif, see *Kitāb Ibn al-Mushrif*
K. al-Īḍāḥ al-shāfī, **II.32.1**
R. fī Ikhtilāf al-maṭāliʿ, App. B, *ad* pp. 485-486
al-Iklīl, II.1.0; **II.1.1**
Īlkhānī Zīj, See *al-Zīj al-Īlkhānī*
R. fī ʿIlm al-falak (?), App. B, *ad* p. 484
R. fī ʿIlm al-mīqāt (?), App. B, *ad* p. 484; App. B, *ad* pp. 485-486; App. B, *ad* p. 487
R. fī ʿIlm al-nujūm (?), App. B, *ad* p. 484
K. al-Inshāʾ fī ʿilm al-jabr wa-l-muqābala, App. A, 3.1

K. al-Jabr wa-l-muqābala, see al-Khwārizmī (author)
Jadāwil fī ʿilm al-falak (?), App. B, *ad* p. 485
al-Jadāwil al-muḥaqqaqa fī ʿilm al-hayʾa (?), App. B, *ad* pp. 484-485
Jadwal fī ʿilm al-falak (?), App. B, *ad* p. 488
al-Jadwal al-thamīn, App. B, *ad* p. 488
Jadwal fī maʿrifat ittifāq al-maṭāliʿ, **II.26.2**
Jawāb al-sāʾil. . . , **II.45.1**
Jawāhir al-ḥisāb, App. A, 3.3

al-K. al-Kabīr, see *al-Kitāb al-kabīr*
al-Kāfī (al-Marrākushī), I, note 16
K. al-Kāfī (al-Ṣardafī), App. A, 1.1
al-Kāfī fi l-farāʾiḍ, App. A, 1.2

Kanz al-thiqāt, **II.47.2**
Kifāyat al-muhtadā, App. A, 1.1
Kitāb Ibn al-Mushrif, **II.12.1**
al-Kitāb al-kabīr (Maᶜshūq al-Qararī), II.2.1

Lubb al-lubab, **App. A, 6.1**
al-Lumᶜa fī ᶜilm al-falak (?), App. B, *ad* p. 484-485

Maᶜārij al-fikr, I.3; I, note 17; II.6.0; **II.6.2**; App. B, *ad* p. 483
K. Mabādiʾ al-ghāyāt, II.18.1
Maᶜdan al-jawāhir, App. B, *ad* p. 488
Majmūᶜ al-zīj (?), App. B, *ad* p. 486
Malḥama tunbaʾ fīhā bimā sayakūn (?), App. B, *ad* p. 488
K. al-Manāzil... (?), **II.2.1**
Manhaj al-ṭullāb, **II.8.2**
Manẓūma fī l-shuhūr al-Rūmīya, II.15.2
al-Maqṣad al-ḥasan, II.34.1
R. fī Maᶜrifat al-awqāt (?), App. B, *ad* pp. 484-485
R. fī Maᶜrifat ittifāq al-maṭāliᶜ, App. B, *ad* pp. 485-486
R. fī Maᶜrifat samt al-qibla, App. B, p. 484; see also *R. fi Samt al-Qibla*
R. fī Maᶜrifat ẓill al-zawāl, App. B, *ad* pp. 485-486
al-Maṭālib al-sunnīya/al-saniya, App. B, *ad* p. 487
Maᶜūnat al-ṭullāb, App. A, 1.1; App. A, 4.1
K. al-Mawāqīt, II.5.1
al-Mazīḥfīya, App. A, 3.2
Miftāḥ al-asrār, **II.27.1**
Miftāḥ al-fāʾid, App. A, 9.2
Mirʾāt al-zamān, I.3; I.4; II.9.0; II.9.2; **II.9.3**; II.9.4; II.18.1; II.38.0
K. Mirʾāt al-zamān, edition of, II.38.2
al-Mudkhal (al-Balkhī), I, note 16
al-Mudkhal al-mukhtaṣar li-Zīj Ibn al-Shāṭir, App. B, *ad* p. 485
Mudkhal al-taᶜlīm, I.3; **II.19.1**
Mufīdat al-sāʾil, **II.41.1**; Addenda, no. 4
Muᶜīn al-ṭālib, **II.8.2**
Mukhtār Zīj, See *al-Zīj al-Mukhtār*
al-Mukhtaṣar fī akhbār al-bashar, II.9, note 2
Mukhtaṣar al-Hindī, App. A, 1.1
al-Mulakhkhaṣ fi l-hayʾa, Addenda, no. 6
Mumtahan Zīj, see *al-Zīj al-Mumtaḥan*
Muntaha l-idrāk, I, note 16
Muntaha l-suʾl, II.6, note 3
Muntaḥal (?) *Zīj*, See *al-Zīj al-Muntaḥal*
al-Muqaddima al-durrīya, **App. A, 5.1**
al-Muqaddima fi l-hisab, **App. A, 3.2**
R. fi l-Muqanṭar (?), App. B, *ad* pp. 485-486
Muṣṭalaḥ Zīj, See *al-Zīj al-Muṣṭalaḥ*
Muẓaffarī Zīj, See *al-Zīj al-Muẓaffarī*
al-R. al-Muẓaffarīya fi l-ᶜamal... bi-l-ṣafīḥa al-jawzaharīya, II.6.4; **II.6.5**

al-Nafḥa al-nadīya, **II.41.1**
Naṣb al-shark, App. B, *ad* p. 488
R. Nihāyat al-idrāk, 1.3; **II.6.4**; II.6, note 3; App. B, *ad* p. 483
Nukhabat al-tuffāḥa, App. A, 2.1

Poem on the lunar mansions, II.24a.1; see also *Manẓūma...* and *Qaṣīda...*

al-Qāmūs, II.32.1
al-Qānūn, I, note 16
Qaṣīda, see also Poem and *Urjūza*
Qaṣīda bāʾīya, App. B, *ad* p. 483
Qaṣīda on the lunar mansions, **II.14.1**; **II.25.1**

Rafᶜ al-ishtibāh, App. B, *ad* p. 488
al-Riyāḍ al-naffāḥa, **App. A, 11.1**
R. fi l-Rubᶜ, see also *R. fi l-Muqanṭar*
R. fi l-Rubᶜ al-mujayyab, App. B, *ad* pp. 484-485; App. B, *ad* pp. 485-486
R. fī... Ruʾyat al-hilāl, App. B, *ad* p. 494

R. fī Samt al-qibla, App. B, *ad* pp. 484-485; see also *R. fī Maᶜrifat samt al-qibla*
Sarāʾir al-hikma, I.3; II.1.0; II.1.3; **II.1.4**
Shāhī Zīj, see *al-Zīj al-Shāhī*
K. al-Shāmil fī dalāʾil al-qibla, **II.26.1**; App. B, *ad* p. 486
Shams al-awān, App. B, *ad* p. 487
K. al-Shams al-Ḥarīrī (?), II.18.1
Shams al-ᶜulūm..., **II.3.1**
Sharḥ Miftāḥ al-fāʾid, App. A, 9.2
Sharḥ Mukhtaṣar al-Khwārizmī, **App. A, 3.1**
al-Sharīda ilā dhikr shuhūr al-Rūm, **II.25.1**
Shāwī Zīj, See *al-Zīj al-Shāwī*
Ṣifat jazīrat al-ᶜArab, II.1.0; **II.1.2**
Sirāj al-tawḥīd..., **II.15.1**
Sulṭānī Zīj, see *Zīj* of Ulugh Beg

al-Tabṣira (Sultan al-Ashraf), I.3; **II.8.1**; II.11.1
al-Tabṣira fī ᶜilm al-ḥisāb, App. A, introd.
al-Tafhīm, I.3; I, note 17
Taiz corpus of tables for timekeeping, **II.9.3**; II.18.1; II.38.2
Takmīl al-shuhūr al-Yazdigirdīya (?), App. B, *ad* p. 486
K. al-Ṭāliᶜ wa-l-maṭāriḥ, **II.1.5**
Taqwīm al-ᶜArabīya al-saᶜīda, **II.48.1**

Taqwīm al-buldān, II.9.0
Taqwīm al-sana, **II.38.1**
Taqwīm ṭawāli^c al-Yaman (?), **II.44.1**
al-Ṭarīqa al-ᶜaẓīma, App. A, 9.1
Ṭarīqat al-ḥussāb, **App. A, 9.1**
Ṭarīqat Jaḥḥāf, **App. A, 9.1**
al-Ṭarīqa al-jalīla, **App. A, 9.1**
K. Tatmīm ᶜamal al-asṭurlāb, **II.13.1**
Ṭaylasān zīj, see *Zīj al-ṭaylasān*
Taysīr al-maṭālib, see *Zīj* of al-Kawāshī
K. al-Tuffāḥa, App. A, 2.1
Tuḥfat al-awān, **II.9.3**
Tuḥfat al-muḥāḍir, II.36.1
Tuḥfat al-rāghib, **II.6.1**; II.6.2

ᶜUqd al-durar, **II.15.1**
Urjūza, see also Poem and *Qaṣīda*
Urjūza fi l-shuhūr al-Rūmīya, II.3.2
Urjūza on the zodiac, II.14.2

al-R. al-Yamanīya fi l-ḥisāb, App. A, 12.1
K. al-Yawāqīt fī ᶜilm al-mawāqīt (al-Aṣbaḥī), **II.5.1**; II.32.1; App. B, *ad* p. 483
al-Yawāqīt fī maᶜrifat al-mawāqīt (Abu l-ᶜUqūl), **II.9.4**
K. al-Yāwāqīt fī maᶜrifat al-mawāqīt (Ibn Jaḥḥāf), **II.39.1**

Zād al-musāfir, see *Zīj* of al-Daylamī
Zahr al-zuhūr, **II.41.1**: **II.47.1**
Zīj, unidentified Abbasid, II.7.1
Zīj of Abu l-ᶜUqūl (≠*al-Zij al-Mukhtār*), **II.9.2**; II.9.4; II.18.1
al-Zīj al-ᶜAlāʾī, II.6.3
al-Zīj al-Balīgh, I, note 16
Zīj of al-Daylamī, I.3; II.7.1; II.20.1; II.23.1; II.28.0; II.28.1; II.29.1; II.36.1
*Zīj*es of al-Fahhād, II.6.3
Zīj of al-Fārisī, see *al-Zīj al-Muẓaffarī*
al-Zij al-Ḥākimī, I.3; II.9.1; II.11.1; II.17.1; II.17, note 1
al-Zīj al-Ḥākimī li-ṭūl Taᶜizz, II.17.1
Zīj of al-Hamdānī, I.3; II.1.0; **II.1.3**
Zīj of Ibn al-Dahhān, II.18.1
Zīj of Ibn Isḥāq, I.3; II.28.1
Zīj of Ibn al-Mushrif, II.12.1
Zīj of Ibn al-Shāṭir, App. B, *ad* p. 485
Zīj of Ibn Yūnus (≠*Ḥākimī Zīj*), II.9.1
al-Zīj al-Īlkhānī, I.4; I.5; II.8.2; II.18.1
al-Zīj al-Jadīd, see *Zīj* of Ibn al-Shāṭir, *Zīj* of Ulugh Beg
Zīj of al-Kaᶜbī, **II.20.1**
Zīj of al-Kawāshī, I.3; II.5a.0; **II.7.1**; II.28.1; II.38.0
Zīj of Kūshyār, see *al-Zīj al-Balīgh*
al-Zīj al-majmūᶜ, **II.35.1**; II.38.0; see also App. B, *ad* p. 486
Zīj of Muḥammad ibn Yaḥyā, II.6.1
al-Zīj al-Mukhtār, I.3; I.4; II.9.0; **II.9.1**; II.9.2; II.9.4; II.18.1; II.28.1; II.41.1
al-Zīj al-Mukhtaṣar, **II.28.1**
al-Zīj al-Mumtaḥan, I.3; I, note 8; II.1.3; II.6.1; II.7.1; II.16.1; II.16, note 1
al-Zīj al-Mumtaḥan al-ᶜarabī, II.6.3; **II.21.1**
al-Zīj al-Mumtaḥan al-khazāʾinī, **II.6.3**
al-Zīj al Muntaḥal (?), II.17.1
al-Zīj al-Muṣṭalaḥ, II.17, note 1; II.18.1
Zīj al-Muthannā, see *Zīj* of al-Muthannā al-Sarḥī
Zīj of al-Muthannā al-Sarḥī, I.4; I, note 20; II.16.1; II.36.0; **II.37.1**; II.41.1
al-Zīj al-Muẓaffarī, I.3; I.4; II.6.0; **II.6.3**; II.21.1; II.35.1; II.36.1; II.41.1
Zīj of al-Najrānī, **II.23.1**; II.28.1
al-Zīj al-Naṣīrī, see *al-Zīj al-Īlkhānī*
Zīj for Sanaa (anonymous), **II.16.1**
Zīj of al-Sarḥī, I.4; I, note 20; II.36.0; **II.36.1**; II.37.1
al-Zīj al-Shāhī, II.17.1; II.17, note 1
al-Zīj al-Shāwī, II.17.1
al-Zīj al-Sulṭānī, see *Zīj* of Ulugh Beg
Zīj for Taiz (anonymous), I.3; **II.17.1**
Zīj al-ṭaylasān, II.17.1
Zīj of al-Ṭūsī, see *Īlkhānī Zīj*
Zīj of Ulugh Beg, I.4; II.38.1
R. al-Ẓill al-mabsūṭ, **II.6.6**
al-R. al-Ẓillīya, II.6.4; **II.6.6**

Index of Modern Authors

Ahlwadt, W., I.3; I, note 14; II.2.0
al-Akwaᶜ (Ismāᶜīl), I note 18
ᶜAlī al-Sharafī, see al-Sharafī
Azzawi, A., I.3; I, note 13; II.6, note 3; II.28, note 1; App. A, introd.; App. A, 3.0; App. A, 3.3; App. A, 8.0

Brockelmann, C., I.3; I, notes 1 and 10; II.6, note 3; II.11a.0; II.14.1; II.15, note 1; App. A, introd.; etc.

Ghul, M., I.3
Glaser, I.3
Goldstein, B. R., I, notes 17 and 28
Griffini, E., I.3; I, note 14; II.23, note 1; II.24.0; Sigla, introd.

al-Ḥātimī (Sanaa astronomer), I, note 20
Heinen, A., I, note 16
al-Ḥibshī, A., App. B, passim

Ismāᶜīl al-Akwaᶜ, see al-Akwaᶜ

Jazāʾirī, T., I, note 12; II.8, notes 3 and 12

Kennedy, E. S., I.3; I, notes 5 and 11

Landberg, I.3
Lee, S., I.3; I, note 12; II.6, note 9
Löfgren, O., I.3; I, note 13; Sigla, introd.

Mayer, L. A., II.6, note 3
Meinecke-Berg, V., I, note 23

Nallino, C., I, notes 10 and 12

Sayyid, A., I, note 18; II.18a
Sergeant, R. B., I.2; I.6G; I, note 3; II.43.1; II.44.1; Addenda, no. 7
Sezgin, F., I, note 1; II.1.4
al-Sharafī (ᶜAlī), I, note 20; II.37, note 1
Suter, H., I.3; I, note 9; II.6, note 3; App. A, introd., etc.

Ullmann, M., II.6, note 3

Varisco, D., I, note 3; Addenda, no. 7

Index of Localities

Localities associated with astronomical activity are listed. Those associated with individuals (places of birth and death) and modern libraries are not.

Aden, II.6.0; II.6.1; II.26.3; II.26, note 1
Alexandria, II.7.1

Bayt al-Faqīh, I, note 20; II.41.1

Cairo, II.13.0
Crete, II.38.0

Egypt, I.1

Fustat, II.2.1

Hadramawt, II.33.1

Ibb, II.28.1
Istanbul, I, note 18

al-Janad, I.2; II.5.1

al-Kawāsha (near Mosul), II.7.1

al-Maḥalla (Egypt), II.38.0
Maragha, II.4; II.17.1
Mecca, I, note 16; II.2.0; II.15.0

Qus, II.7.1

Samarqand, II.38.1
Sanaa, I, note 20; II.2.1; II.16.1; II.24.0; II.29.1; II.3.1; II.38.0; II.38.1; II.40.1; II.44.1; II.48.1
Syria, I.1

Taiz, I.4; II.7.1; II.8.2; II.9; II.17.1; II.26.3; II.44.1; II.48.2
Tunis, II.28.1

Zabid, II.12.1; II.33.0
Zabid, algebra invented in, App. A, introd.
Zofar, I.5

General Subject Index

Abbasid astronomers, I.2; I.3; II.1.3
agriculture, I.2, etc.
algebra, App. A, passim
algebra, invented in Zabid, App. A, introd.
almanacs, I.2; I.4; I.6G; etc. See also Addenda, no. 7
Arab League, see Cairo
arithmetic, App. A, passim
arithmetic of inheritance, App. A, passim
Armenian script, II.17.1
armillary sphere, II.8.2
ascensions, II.26.2, etc.
astrolabe, al-Ashraf's, I.2; I.3; II.8.2; II.8, note 12
astrolabe tables, I.4; II.8.2; II.8, note 7; II.13.1
astrolabes, I.4; II.8.2
astrology, I.6F; I, note 21; II.1.5; II.6.4; II.6.7; II.8.1; II.30.1; etc.
astrologers, modern Yemeni, I, note 20
astronomy, mathematical, I.2, etc. See also folk astronomy
astronomers, modern Yemeni, I, note 20
automata, I.5; I, note 22

Byzantine astronomers, II.6.3

Cairo, Arab League Institute for Arabic Manuscripts, Addenda, no. 5
Cairo corpus of tables for timekeeping, I.4; II.9.3; II.13, note 1
Cairo, manuscript libraries, I.2; I.3
calendar, Hijra, II.21.1, etc.
calendar, Persian, II.6.3; II.18.1; II.21.1
calendar, Syrian, II.26.3; II.33.1
calendar conversion, device for, II.46
calendrical tables, I.4; I.6H, etc.; I.41.1
celestial globe, II.38.0
compass, magnetic, I.2; I.3; I.4; II.8.2; II.8, note 11

Damascus corpus of tables for timekeeping, II.9.3
dates of events in Yemen, II.22.1
dating, problems of, II.6.4
divination, I, note 21

eclipses, I.5
eclipse computer, II.6.5; II.6, note 12
Egypt, astronomy in, I.1
ephemerides, I.4; I.6D, etc.; I, note 28; II.11.1; II.38.1
equatoria, II.6, note 12

folk astronomy, I.2; I.6A, etc.

Geniza, I, note 28
gnomon in Janad, I.2
gnomons, I.2; I.3, etc.; II.1.2; II.8.1

Hebrew, Arabic texts written in, I.3; I, note 17
historical events, list of, II.22.1
horoscopes, II.11.1; II.22.1; II.30.1; etc.

ijāzas for astrolabes, II.8.2
Indians (value of the obliquity), I.4; II.8.1
inheritance, App. A, passim
instruments, I.6E, etc.
Iraq, see Abbasid astronomers
Istanbul, Yemeni manuscripts in, I, note 18

Kaᶜba, orientation of, II.6.1; II.6, note 6

lunar crescent visibility, see moon

magic, I.4; I, note 21
Milan, Biblioteca Ambrosiana, I.2
microfilms of Yemeni manuscripts, I.3; Addenda, no. 5
mīqātīya, I.5
moon: lunar crescent visibility, II.9, note 4; II.11.1; II.26.2; App. B, *ad* p. 494;
lunar equation tables, II.9, note 4; II.18, note 3;
lunar mansions, I.2; I, note 2; etc.

New York, Metropolitan Museum of Art, I.2
non-Yemeni works, I.2; I, note 16; Addenda, no. 6

obliquity, determined by observation, II.7.1
observations, II.7.1
observations, "new", II.38.1
Ottoman convention for reckoning time, II.47.2; see also II.40.1
Ottomans, I, note 18
Oxford, eclipse computer, II.6.5

Pahlavi, II.6.7
planetary equation tables, unusual, II.29.1

prayer, times of, I, note 27; II.2.1
prayer-tables for Sanaa, modern, II.48.1
prayer-tables for Taiz, modern, II.48.2
prayer-tables for the Yemen, medieval, II.33.1; II.40.1

qibla, II.2.1; II.6.1; II.6, note 5; II.8.2; App. B, *ad* p. 484; App. B, *ad* pp. 484-485
qibla table, II.7.1; II.7, note 3
quadrants, App. B, *ad* pp. 484-485 and *ad* pp. 485-486

rainbow, II.11.1
Rasulid treasury, II.6.2; II.6.4
Rasulids, I.3; I, note 19; etc.

ṣafīḥa jawzaharīya, II.6.5
shadow-schemes, II.6, note 5
shadows, measurement of, II.2.1
shamaʿdān, I.5
sphere, see armillary sphere, celestial globe

sun: solar eclipse tables, II.16, note 1; solar halos, I.5
sundial tables, II.8.2; II.8, note 9
sundials, I.4; II.6.6; II.8.2
Syria, astronomy in, I.1

tables for timekeeping, I.1; I.6; I, note 6; etc.
tables: see also calendrical tables, moon, prayer-tables, qibla table, sun, *zīj*es
Taiz corpus of tables for timekeeping, II.9.3; II.18.1
time, *ʿarabī*, II.48.1; II.48.2
time, *ifrankī*, II.48.2
times of prayer, see prayer

Yemen, manuscript libraries in, I.2; I, note 18
Yemen, research facilities in, I, note 15

Zaydīs, II.25.0; II.34.1
*zīj*es, I.1; I, note 5; I.6B, etc.
*zīj*es, topics of, II.6.2

Index of Terrestrial Latitudes

The reader should bear in mind that I have not extracted the coordinates for Yemeni cities in medieval Yemeni *zījes*.

11;0°	II.7.1
13;0°	II.6.1; II.8.1; II.8.2; II.9.0; II.9.1
13;10°	II.9.0
13;37°	II.8.1, II.8.2; II.9.2; II.17.1
13;40°	II.7.1; II.9.1; II.9.2; II.9.3; II.22.1
13;43°	II.8.1
14;0°	II.8.1; II.8.2; II.9.1; II.12.1
14;5°	II.9.0
14;20°	II.17.1
14;30°	II.6.3; II.8.2; II.9.1; II.24.0
15;0°	II.8.2; II.40.1
15;30°	II.24
21;0°	II.8.2
24;0°	II.8.2

Index of Values of the Obliquity of the Ecliptic

24;0°	II.8.1; II.24
23;51°	II.8.2
23;35°	II.8.2; II.9.1; II.9.3; II.22.1; II.24
23;33°	II.7.1; II.8.2
23;32,50°	II.7.1
23;30°	II.8.1; II.8.2

ADDENDA

(1) *ad* 34.1, no. 12: Additional copies of the tables entitled *Bulghat al-muqtāt* are MSS Cairo Dar al-Kutub K 3769,2; London B.L. Supp. 773,4; Vatican V.1086$_1$.

(2) MS SP, located after this study was completed, contains an anonymous almanac for the Syrian months, probably of Yemeni origin. This work merits detailed investigation.

(3) To the list of Yemeni mathematicians add al-Faḍl b. Abī Saʿd al-Uṣayfirī (*Brockelmann*, I, p. 510, and SI, p. 702); Ibrāhīm b. Muḥammad al-Bawsī (*Brockelmann*, SI, p. 702, and SII, p. 242); Aḥmad b. Yaḥyā al-Ṣaʿdī al-Dawwārī (*Brockelmann*, SII, p. 559); Aḥmad b. Muḥammad al-Khālidī (*Brockelmann*, I, p. 510 and SI, p. 702).

(4) *ad* 41.1: Another copy of the tables of Yaḥya b. Muḥsin entitled *Mufīdat al-sāʾil* appears to be in MS LQ (104 pp. ??).

(5) A list of manuscripts microfilmed in the Yemen by the Arab League Institute for Arabic Manuscripts has been published in *Revue de l'Institut des Manuscrits Arabes*, 24:1 (1978).

(6) MS Cairo Dār al-Kutub *hayʾa* 69 (18 fols., ca. 900H) is a Yemeni copy of the early-thirteenth-century *Mulakhkhaṣ fī l-hayʾa* by Maḥmūd b. Muḥammad b. ʿUmar al-Jaghmīnī. The manuscript is copied in the same hand as MS TG. The copyist appears to be innocent of any knowledge of astronomy.

(7) My friend Dr. Daniel Varisco has kindly drawn my attention to R. B. Serjeant and H. A. al-ʿAmri, "A Yemeni Agricultural Poem," in W. al-Qāḍī, ed., *Studia Arabica et Islamica: Festschrift for Iḥsān ʿAbbās*, Beirut: American University of Beirut Press, 1981, pp. 407-427. In this study Serjeant states that the manuscript from which he has extracted the poem also contains a medieval almanac and information on eclipses.

(8) Two more copies of al-Shayzarī's *ʿĀdāt al-nujūm* are contained in MSS Milan Ambrosiana D277 and D487.

CAPTIONS

Plate 1. An extract from MS BN (= Berlin Ahlwardt 5720 = Mq. 733) of the corpus of tables for timekeeping compiled by Abu l-ʿUqūl. This particular set of tables displays the ecliptic longitude of the ascendant (point of the ecliptic instantaneously rising over the horizon) for each degree of solar longitude when the solar altitude is 3°. Similar sets of tables are given for each degree of solar altitude up to the maximum for Taiz. Values are given for altitudes in the east and in the west. Thus one measures the solar altitude at any time with an instrument and the table displays the corresponding longitude of the ascendant—an astrologer's dream! (No indication is given about the practical use of these tables in Taiz which is nestled at the foot of a mountain slope facing north!) No such tables are known to have been compiled for any other locality in the Muslim world. Other tables in this corpus include tables for reckoning time from solar altitudes and from stellar altitudes. Abu l-ʿUqūl's tables are more extensive than any other medieval set of tables for timekeeping, containing a grand total of about 100,000 entries.

Plate 2. An extract from the anonymous almanac for Taiz, 727 Hijra (= 1326-27) preserved in MS TG (= Cairo DM 817). These two pages serve the month of Rabīʿ I, each line of entries being for a day of the lunar month. In the right-hand margin information is given about festivals and agricultural activities, and then five columns give the day of the month, the corresponding day of the week, and the date in the Syrian, Coptic, and Persian months. The next columns give the midday positions of the moon, sun, Saturn, Jupiter, Mars, Venus, Mercury, and the lunar node, in degrees and minutes of each zodiacal sign. The last two columns display the number of hours of daylight and the solar meridian altitude. (It was from the meridian altitudes that the almanac was associated with the city of Taiz.) On the left-hand page there is a written statement about the position of the moon in the zodiacal signs and its relationship to the positions of the sun and other planets, together with a rather vague and general prognostication: *ṣāliḥ* or *fāsid* or *mutawassiṭ*, that is "good news" or "bad news" or "either". At the top of the right-hand page is a calculation of the exact time of the solar-lunar conjunction which marks the beginning of the astronomical lunar month, and at the top of the left-hand page is a statement of the evening on which the lunar crescent will be seen (in this case on day 2 of the month), together with the results of the calculations on which this prediction was based. See also Plate 3.

Plate 3. An extract from the anonymous almanac for Taiz, 727 Hijra. See already Plate 2. Whereas the almanac and ephemerides are arranged for the months of the Muslim lunar year, additional information on the corresponding solar months is given at the end of the almanac. This page shows four horoscopes for the moments of the equinoxes and solstices, giving the precise positions of the twelve astrological houses on each occasion. The text states that the calculations were made with the *Ḥākimī Zīj*, that is, the major work of the tenth-century Egyptian astronomer Ibn Yūnus, by which we may assume a Yemeni recension of that *Zīj* since the original contains planetary tables for Cairo and no ascension tables specifically for Taiz. Such a Yemeni recension of Ibn Yūnus' *Zīj* exists in MS PA (= Paris Bibliothèque Nationale ar. 2523).

Plate 4. An extract from the solar and lunar tables in MS YA (= Cairo DM 400) of the *Zīj* of the tenth-century Iranian astronomer Kūshyār ibn Labbān. The right-hand page shows part of the table of the solar equation with marginalia by a later Yemeni astronomer on how the table is to be used for a specific astrological purpose. The left-hand page shows the table for the mean motion of the moon, with a statement (above the upper left-hand corner of the frame of the table) of the correction to be applied to the mean lunar position to account for the longitude difference between Iran and the Yemen. The tables are based on the Persian calendar, but at least for a medieval astronomer, this would not make them especially difficult to use.

Plate 5. The back of the astrolabe made in 690 Hijra (= 1291) by ʿUmar, the son of the Rasulid Sultan al-Muẓaffar, who in 1295 became Sultan himself and was called al-Ashraf. The inscriptions are the same as those in the illustrations in al-Ashraf's treatise on the astrolabe: see Plate 6. The planetary symbols are identified by the names of the planets in the space at the center of the instrument. All of this information could have been compressed into one half or one quarter of the available space, to make room for the more useful trigonometric grids and calendrical scales that are more commonly found on medieval astrolabes. On the rim of the lower right quadrant is a simple scale indicating the shadows of an object 12 units long cast by the sun when it has an altitude measured in the upper left quadrant. The inscription on the rim of the lower left quadrant translates:

> "This astrolabe is the work of ʿUmar ibn Yūsuf ibn ʿUmar ibn ʿAlī ibn Rasūl al-Muẓaffarī with his own hand and to the best of his ability in the year 690 Hijra."

The alidade is unusual in that it bears a sighting tube. Celestial altitudes are to be measured on the scales on the rim of the upper quadrants. The front of the astrolabe is standard and the instrument contains three plates that are original, serving the latitudes of Aden, Taiz, Sanaa, North Yemen, Mecca and Medina, as well as a fourth plate which is taken from another instrument.

On a scale of A to D al-Ashraf's astrolabe merits at most a B, but his teachers gave him an A: their comments on this astrolabe and three others made by him are appended to the Cairo manuscript of al-Ashraf's treatise.

Plate 6. An illustration of the astrological information that was occasionally recorded on the backs of astrolabes, as well as in treatises on astrology, taken from MS TA (= Cairo TR 105) of the treatise on instruments by the Rasulid Sultan al-Ashraf. This diagram is of interest because it displays precisely the information engraved on the back of one of al-Ashraf's astrolabes now preserved in the Metropolitan Museum of Art in New York: see Plate 5.

The altitude scales for 0° to 90° are marked in 5° intervals and 1° subintervals on the rim of the upper two quadrants. The other markings are arranged according to the zodiacal signs starting with Aries, clockwise from the top. The names of the signs are written in Arabic, but the other information is written using planetary symbols which are explained beneath the diagram. There are three main scales of symbols. The outer scale marks the "terms" of the signs, five areas of unequal size for each sign, whose length in degrees is indicated by the number below the symbol of the planet associated with that term. The middle scale displays the "lords of the faces", the faces being one-third divisions of each sign. The inner scale displays the planets associated with the triangles or triplicities. Each triplicity consists of three signs each separated from the next by three others, and each is associated with one of the four elements: fire, earth, air and water. For each triplicity there is a "lord of the day", and a "lord of the night", and a "companion." For each zodiacal sign these three planets are shown on the scale. All of these concepts were inherited by the Muslims from Hellenistic astrology.

al-Ashraf's treatise also contains tables of coordinates for marking the curves on the plates of his astrolabe, as well as a discussion of horizontal sundials and tables for constructing them and an appendix on the magnetic compass: see Plate 7.

Plate 7. An extract from MS TA (= Cairo TR 105) of the treatise on instruments by the Rasulid Sultan al-Ashraf. The illustration displays a magnetic compass and the text represents the earliest description of the compass in an Arabic scientific treatise. However, al-Ashraf makes no claim to be the first to write about the compass and it is certain that he was not. There is no mention of the fact that the compass might not point due north-south, but then we have no information on the amount of magnetic variation in the Yemen at the time the treatise was written. At the bottom of the ring are marked the qiblas of Taiz and Aden at 20° west and east of north respectively: no explanation of this absurdity is presented in the text.

Plate 8. An extract from MS ME (= Milan Ambrosiana Sup. 73 = Griffini 37) of the treatise on folk astronomy by the Adeni astronomer al-Fārisī. The extract deals with the determination of the qibla by non-technical means, by the winds and the stars. The first section deals with the definitions of the directions from which the four cardinal winds blow, these directions defining the alignments of the rectangular base of the Kaʿba. The second section discusses the alignments of the sides of the Kaʿba in terms of the directions of the rising sun and setting moon at the equinoxes and solstices. Recent investigations of aerial maps of the Kaʿba and its environs have established that the major axis is aligned towards the local rising point of the star Canopus, as mentioned in the first text. Also, the minor axis is roughly (within 5°) solstitially aligned, so that, for example, the north-east wall with the door faces the general direction of the rising sun at the summer solstice. However, it is now established that the minor axis is precisely (within 1°) aligned with the southernmost setting point of the moon at the winter solstice. The possible significance of this lunar alignment, which is not specifically mentioned in any of the medieval texts dealing with the Kaʿba and the winds, is still under investigation. For more information see *Hawkins & King*.

Plate 9. In his treatise on folk astronomy (see Plate 9), al-Fārisī presents a description of the twelve geographical areas about the Kaʿba and states how the qibla in each area can be determined by the stars. He then presents the two diagrams shown here of a world divided into twelve sectors. Yet the information recorded in the text and in each of these two diagrams is different. In other such qibla diagrams in non-Yemeni sources, yet other schemes are proposed. The notion of a sacred geography—a world centered on the Kaʿba—has been completely overlooked by all those who have written on Islamic geography. The various schemes described and/or illustrated in the known sources, which presently number twenty, are currently under investigation.

Plate 10. The text of the poem on the Syrian months by the eleventh-century scholar Nashwān al-Ḥimyarī as it appears in MS YK (= Cairo DJ 705,3). The poem gives information about celestial phenomena and agricultural activities, of the kind which is presented in greater detail in Yemeni almanacs. This primitive astronomical folklore is practiced by Yemeni farmers to this day.

PLATES

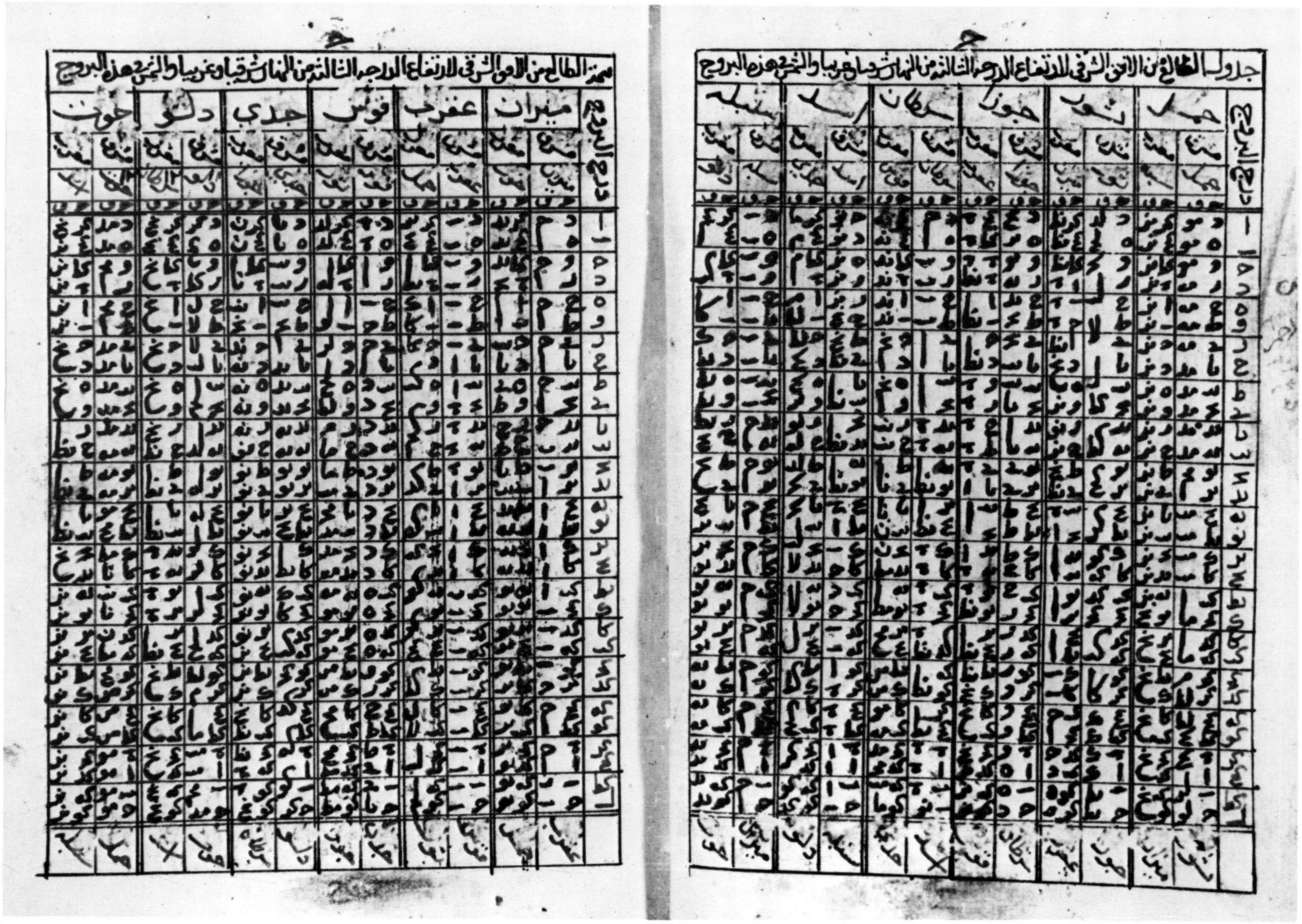

Plate 1

An extract from the corpus of tables for timekeeping compiled in Taiz ca. 1300 by Abu l-ᶜUqūl.
[*Staatsbibliothek, Preussischer Kulturbesitz, Orientabteilung, Shelf No. Mq. 733*]

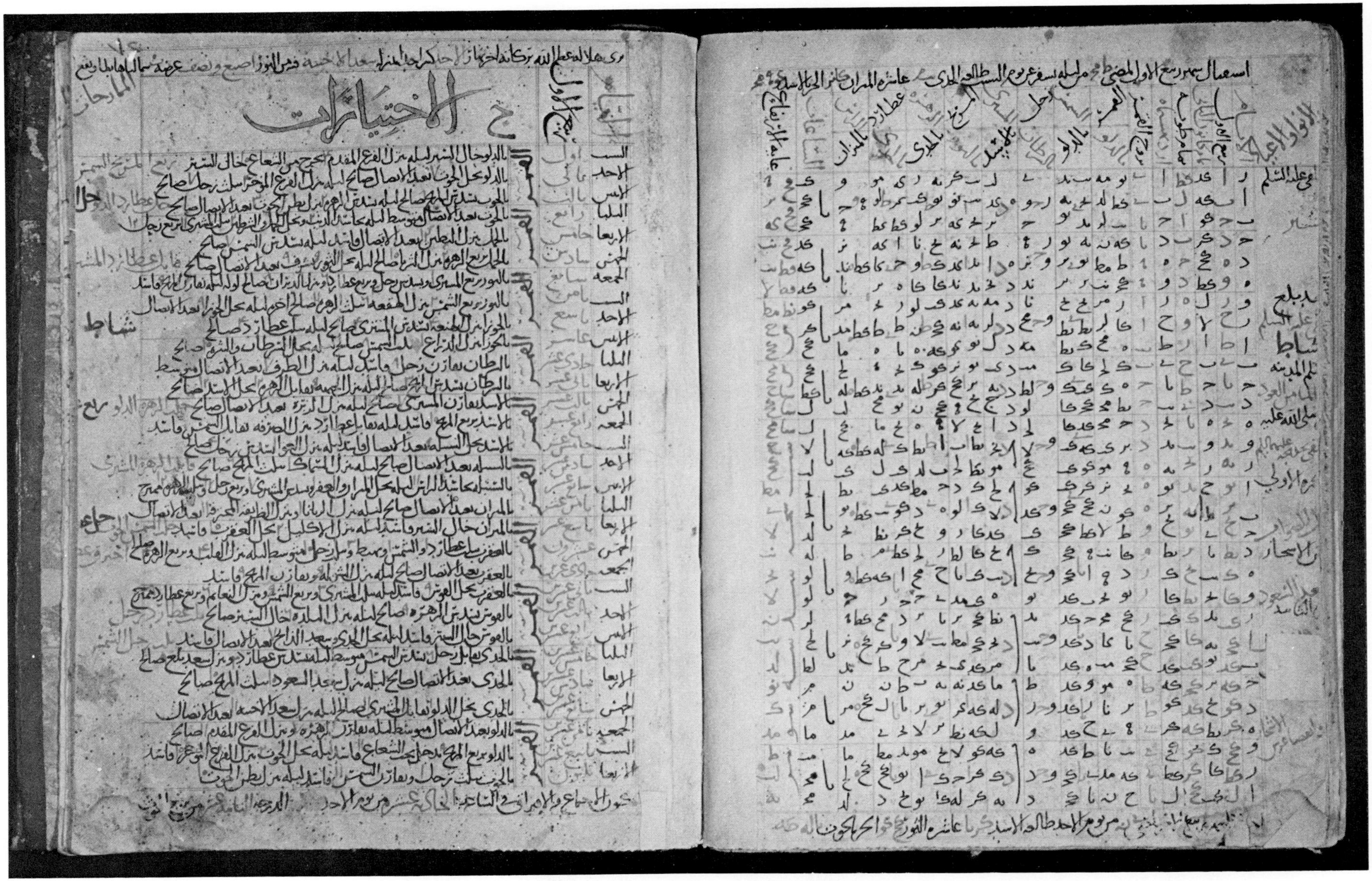

Plate 2

An extract from a fourteenth-century almanac for Taiz
displaying solar, lunar, and planetary positions for each day of one month of a specific year.

الحادي عشر الاسد
التاسع الجوزا
العاشر السرطان
الطالع الميزان
السابع الحمل
الرابع الجدي
الثالث القوس
الخامس الدلو
الحادي عشر السرطان
العاشر الجوزا
الطالع السنبلة
السابع الحوت
الرابع القوس
الثالث العقرب
الخامس الجدي
الحادي عشر الميزان
التاسع الاسد
العاشر السنبلة
الطالع القوس
السابع الجوزا
الرابع الحوت
الخامس الحمل
الحادي عشر الحمل
العاشر الحوت
الطالع الجوزا
السابع القوس
الرابع السنبلة
الثالث الاسد

Plate 3

Astrological horoscopes contained in the almanac illustrated in Plate 2.

Plate 4

An extract from a Yemeni copy of a tenth-century Iranian astronomical handbook.

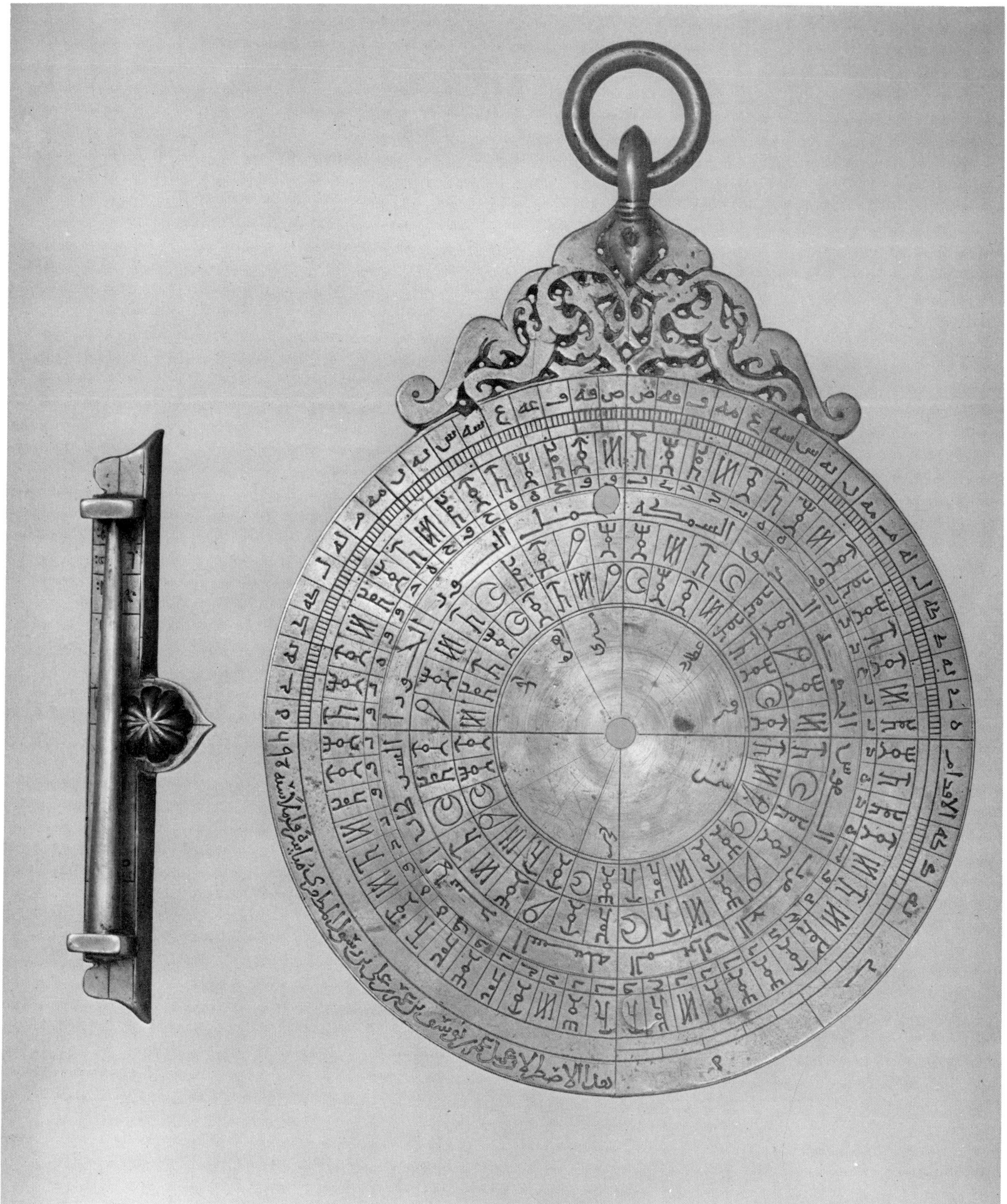

Plate 5

The back of the Sultan al-Ashraf's astrolabe preserved in the Metropolitan Museum of Art in New York. See also Plate 6.
[*The Metropolitan Museum of Art, Bequest of Edward C. Moore, 1891.* [*91.1.535*].]

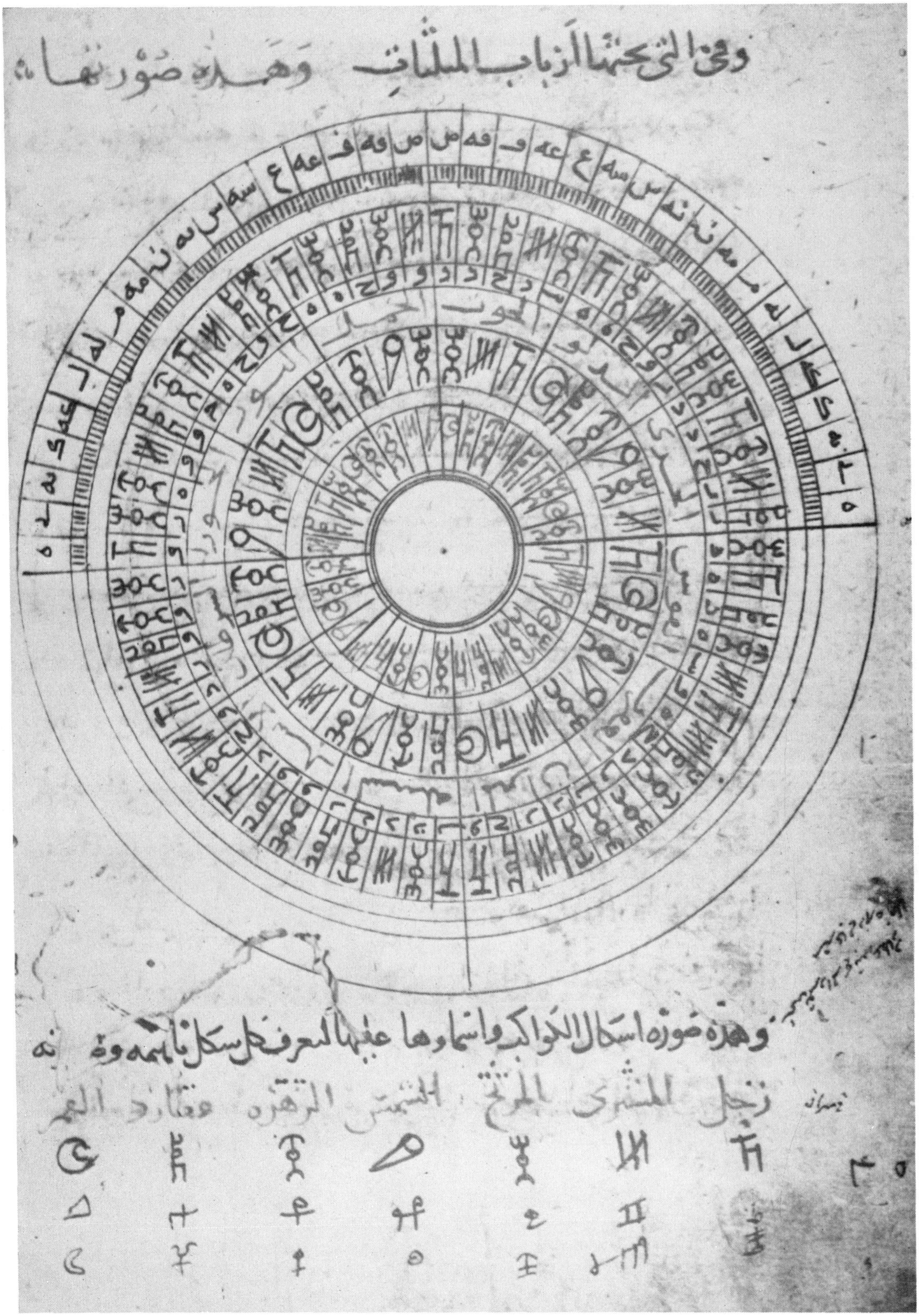

Plate 6

The design for the back of an astrolabe illustrated in al-Ashraf's treatise on astrolabe construction. See also Plate 5.

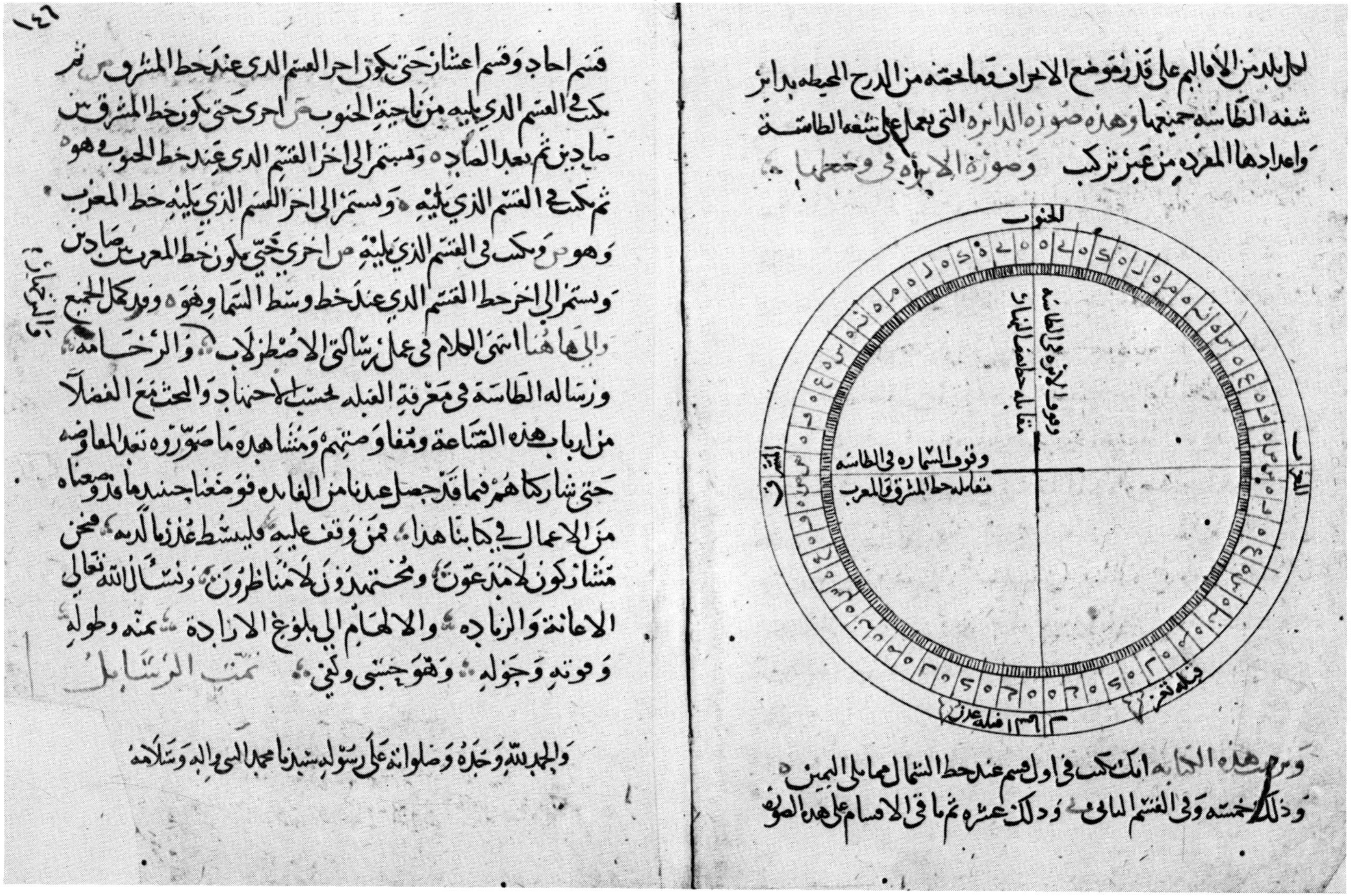

Plate 7

The section on the magnetic compass in the Sultan al-Ashraf's treatise on instrument construction.

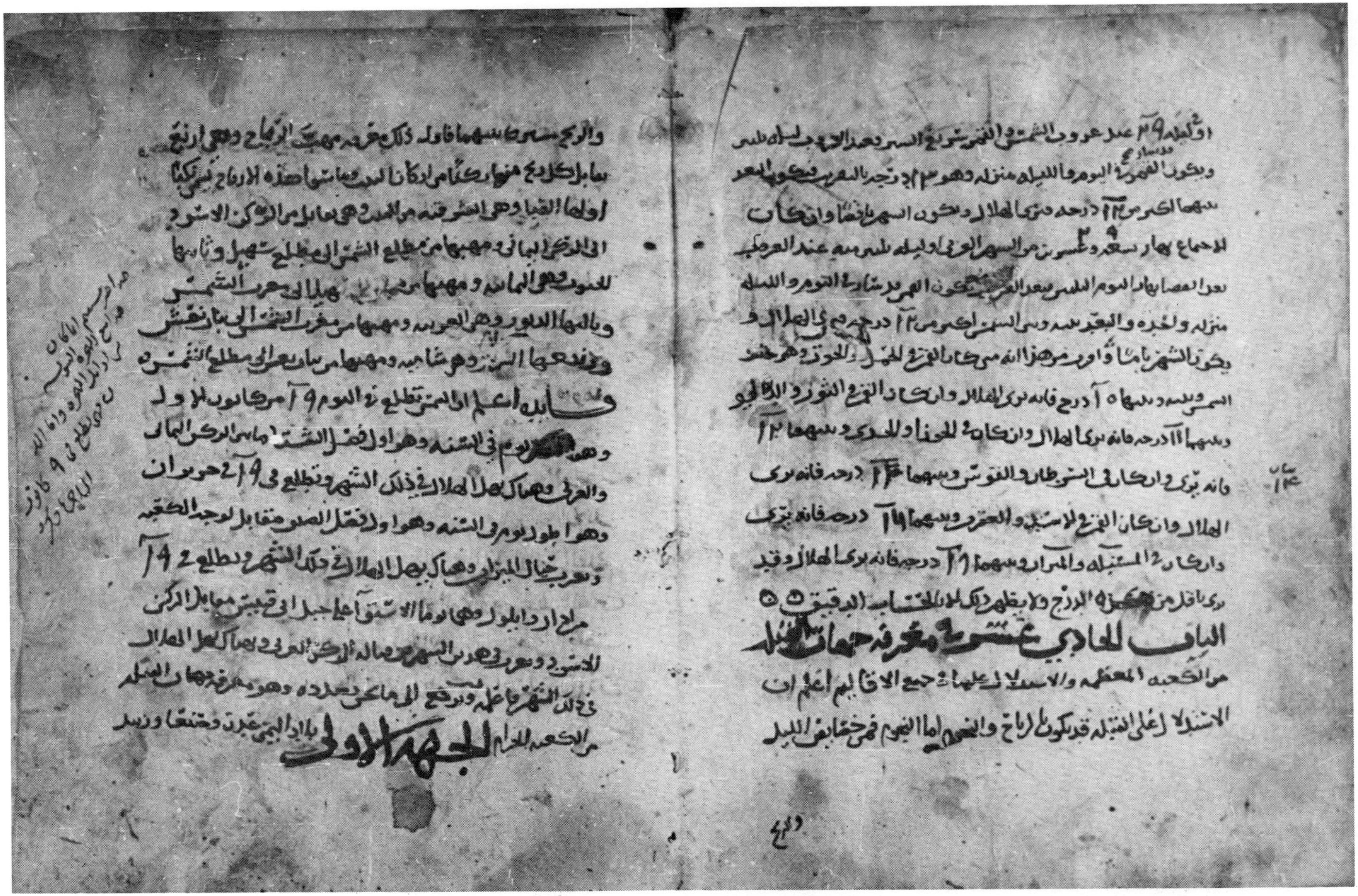

Plate 8

al-Fārisī's text on the orientation of the Kaᶜba and its association with the winds.

Plate 9

The world arranged in sectors about the Kaᶜba, as displayed in al-Fārisī's treatise on folk astronomy.

بسم الله الرحمن الرحيم هذه ارجوزة نشوان الحميري المشهورة

تشرين اول

تشرين ثاني

كانون اول

كانون ثاني

شباط

اذار

نيسان

ايار

حزيران

تموز

آب

ايلول

Plate 10

An eleventh-century Yemeni poem on the Syrian months and associated celestial phenomena and agricultural activities.